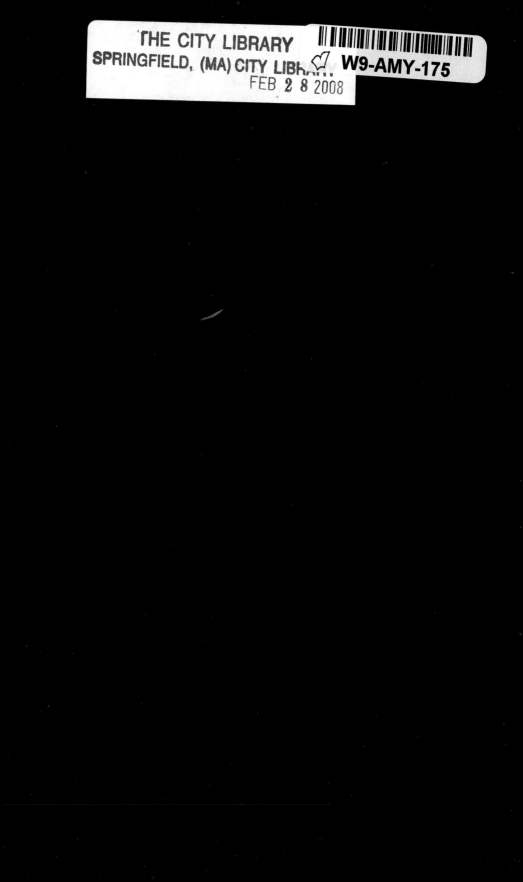

BEYOND THE ZONULES OF ZINN

BEYOND
THE ZONULES
OF ZINN

A FANTASTIC JOURNEY
THROUGH YOUR BRAIN

David Bainbridge

HARVARD UNIVERSITY PRESS

Cambridge, Massachusetts
London, England
2008

Library of Congress Cataloging-in-Publication Data
Bainbridge, David.
Beyond the zonules of Zinn : a fantastic journey through
your brain / David Bainbridge.
p.   cm.
Includes index.
ISBN-13: 978-0-674-02610-0 (cloth: alk. paper)
ISBN-10: 0-674-02610-1 (cloth: alk. paper)
1. Neuroanatomy—Popular works.   I. Title.
QM451.B35 2007
611′.8—dc22   2007021595

For Rose . . .

. . . and those baby blue zonules

CONTENTS

III. Where All the Mind May Be Found?
      The cortex

Hear and attend and listen; for this befell and
behappened and became and was, O my Best Beloved,
when the tame animals were wild.

—Rudyard Kipling

PROLOGUE

Twenty weeks and time for a scan. We had been through this whole process twice before, but in the summer of 2003 we reached that stage once again. In the United Kingdom most maternity hospitals carry out a fetal ultrasound scan around halfway through pregnancy—at about twenty weeks. Some babies are also scanned earlier if there is uncertainty about the date of conception and some are scanned throughout pregnancy if there is particular reason to worry.

Our baby was due for a standard twenty-week "anomaly scan" to check for any major developmental problems or sluggish growth. Yet many parents—and it is easy to fall into this trap—seem to view the scan as a spectator event, an exciting chance to see their child for the first time. The mood in these ultrasonography waiting rooms is often cheerful, expectant, almost celebratory in a quiet way, especially considering the crushing news that must sometimes be delivered to expectant parents.

As with my previous two encounters with this situation, I did not really know how much to watch of the scan itself. Many hospitals in this country have a policy of not telling parents the sex of their baby, and Michelle and I always agreed that we did not want to know. After all, why spoil a surprise? Yet as a vet, I am used to carrying out ultrasound scans on animals, and so I can usually work out roughly what is going on in a human pregnancy scan. And as many parents among you will already realize, "roughly" is often more than sufficient to work out your baby's sex.

As soon as the ultrasound probe plopped onto my wife's belly, a perfect outline of an evenly rounded baby's head sprang onto the screen. A quick flick of the ultrasonographer's hand and a reassuringly complete backbone rocked to and fro before our eyes. It may sound blasé, but this is the point at which I slipped into gawping spectator mode. Two of the commonest abnormalities that the scan is meant to discover are anencephaly and spina bifida, in which either the brain or the spinal cord fails to seal up properly. As I will explain later, these organs form together as a simple tubular structure early in embryonic development. Both conditions are big problems for a baby, yet they are surprisingly common. No one seems to like to mention that anencephaly, in which the top of the skull and most of the brain simply do not form, occurs in perhaps one or two in every thousand human pregnancies. Spina bifida is also relatively common—in a field in which a rate of one in a thousand pregnancies is considered common—but it is a more variable creature than anencephaly, ranging from barely detectable to severely disabling. Both these abnormalities are fairly easy to see on a scan, hence my tendency to relax once they have been ruled out. Somehow, detailed measurement of the length of a thighbone does not seem to hold so much potential as a harbinger of doom.

With the brain and spinal cord examined I would have loved to stare at the scan, calmly soaking in the first view of our baby, but I could not relax fully. Some ultrasonographers have a cavalier attitude to exposing babies' genitals, and whenever the scan crept down toward the nether regions I found myself averting my gaze in a strangely prudish manner. After accidentally diagnosing the sex of both our older children at their scans, I was determined not to be similarly overinformed this time. And yet I failed, and so did my wife. A cruelly aggressive flick of the ultrasound probe and our baby's one great secret was suddenly out—an image that, once seen, could not be ignored. At that moment we both knew, and we each knew that the other knew. In the world of ultrasound sex diagnosis, absence of evidence really is evidence of absence. A baby girl, in other words.

With that pathetic failure under my belt, I settled back to watch the rest of the grainy, black and white show. With nothing specific to look at, or to avoid looking at, I suddenly realized that this scan was very different from the scans of our first two children. Perhaps funding for the British National Health Service is not as limited as we are led to believe, for I think the hospital must have bought a new machine. The quality of the ultrasound images was far better than anything I had seen before. The detail was amazing, so amazing in fact that it almost seemed improper to be peering into our future child in this way. I always wonder if my aversion to looking at the inner workings of people is made worse because I am a vet. Over the years I have had all the squeamishness knocked out of me—I teach anatomy after all, so I am completely immune to gore. Yet I am sure this means that when I am suddenly confronted by the inside of a person, completely unfettered by any sense of visceral disgust, I am unusually sensitive to the humanity of the being exposed to me.

So to see our own future baby, child, friend, and confidante repeatedly sliced and sectioned like this was disturbing and compelling in equal measure. I did watch, but maybe a little from the corner of my eye. And one image that flicked past has lodged itself in my mind ever since, an epiphany of how we all work.

When waves of ultrasound travel through the eye, whether of a baby or an adult, there is not much for them to bounce off, and it is for this reason that an eye looks like an eerie black globe suspended in the grey mishmash of the head. Yet eyes are not entirely featureless on a scan, because hanging within the eye itself is a small dollop of reflective whiteness, the lens. The job of your lenses is to allow you to focus on objects at different distances, and to do this they must be optically denser than the eyeball fluid that surrounds them. Just as a glass lens in air can focus an image onto a sheet of paper, so the lenses in your eyeball fluid can focus the world onto the light-sensitive backs of your eyes. But unlike their crude, artificial imitators, human lenses can change their shape—or rather be changed in shape—and it is this property that allows our gaze to rest comfortably on flowers held in our hands as well as on trees on the horizon. At rest, the human lens is a chubby, almost spherical shape—a slightly flattened globe that sharply bends light rays from nearby objects onto the back of the eye. To make distant objects snap into focus the lens must be flattened and it is this active process in which we are interested here. Although you may not believe it, the story of the lens is an exciting one, a window on the world of how we form and whence we came, but it is a story that will have to wait until later.

For now we are more interested in the tiny machinery that flattens and distorts the lens. Even in a human baby-to-be, whose blurred visual world is only a dull, featureless glow even when its mother exposes her belly to bright sunlight, that machinery is already in place. It was this machinery that I spied in my future baby's little eye. Radiating out from the edge of each lens are hundreds of tiny fibers attaching the lens to a serrated muscular ring. When the muscles in this ring contract, they tug on the minuscule fibers, and it is this tug that flattens the lens and allows you to see into the distance. These fibers really are very small. I know from experience that they are so tiny and fragile that it is easy to overlook or even damage them. Yet tiny as they are, these little strands have a name filled with portent. They are the zonules of Zinn.

Ever since I first saw that name in a school book perhaps twenty-five years ago, it has gripped me. It is like a name from an ancient map, from a souk, from another galaxy. It would be a great name for a psychedelic rock group. As far as I am concerned, with a name like that it hardly matters what the zonules actually do. This is what struck me as I saw them on the screen. The zonules have it all—not only are they the pinnacle of miniature biological engineering, but we have also given them a most alluring moniker. After all, when we think of things in the world around us, we do not just think of the things themselves—we also simultaneously think of the names we have given them. We compulsively stamp every object, being, and concept with a label from which it then cannot be separated. We color everything with the name we give it.

If science is supposed to be a cool, objective, balanced approach to the world, why have we reserved some of our most creative and powerful names for scientific discoveries and concepts? If we distort everything we name, then why throughout history have we gone out of our way to distort our scientific worldview? I will show you in this book that the human brain, and the way it perceives the world—

courtesy of the zonules of Zinn, of course—is at the very center of this wonderful paradox. The names we have given bits of the brain are the strangest of all. The brain is replete with aqueducts, breasts, hillocks, knees, seahorses, and even a tender mother.

And the strangeness of all these names is a clue to the central theme of this book. There is a very good reason why the parts of the brain have such weird names: most of the time we have been thinking about the brain we have had little idea what any of its parts actually do. Over the last fifty centuries anatomists have meticulously drawn a map of the internal geography of the brain, but only in the last century or so have we started to understand the functions of all the regions on that map. For most of recorded human history, all we understood about the brain was structure, not function. And by the time we started to study the function of the brain, its constituent parts had already been given glorious, whimsical, naïve names, which have now irrevocably stuck.

My story of the brain is all about structure, because that was all we knew for all those centuries. It is also the only thing we really know for sure today. And all those centuries and all that certainty also mean that this is the simplest way to understand the brain. When you realize that the brain has a geography as real as that of any continent in an atlas, you will see why this is the easiest way to understand how it works. All the wonderful things the brain does must take place somewhere, and it is this sense of place that dominates this book. In fact, I will argue that we cannot understand any phenomenon in the brain until we have first discovered where it occurs. So this is why this book takes the form of a geographical tour of your nervous system, from the small of your back to the top of your head.

This focus on structure will also lead us on to another exciting perspective from which to view that thinking machine in our heads. All around us are myriad animals with brains designed along the same lines as ours, but adapted to very different lifestyles—swimming through coral, swooping on mice, rooting for acorns. We have thousands of cousin-species with which to compare our brains, and we even have some fossil evidence of how our brains changed into what they are today. The great gift of evolutionary biology is that it allows us to look back in time to see where we came from, and the story of our acquisition of a brain is just as intriguing as the brain we have ended up with. When we look at the brain in this archaeological way (in the second part of the book), I think you will be very surprised by some of our discoveries.

In the third and final part of the book, our concentration on brain structure will also give us a novel approach to some of the greatest mysteries of the brain. We will see how thought, emotion, and consciousness can be approached as engineering problems. The realization that they must occur somewhere makes understanding them simpler—they are phenomena that can be localized, just like the simpler functions of the brain described earlier in the book. But of course there is something spookier about them. When we study the brain, we are doing something unusually self-referential. We are studying the very organ we use to do the studying. We are also studying the part of us that holds our gaze, as the iris and zonules are themselves just specialized outgrowths of the brain. And when we give our fantastic names to parts of the brain, we are coming to terms with naming the physical elements of our mind. Indeed, are we sometimes naming the bits of the brain actually responsible for the arcane and distorting process of naming itself?

This book is my story of the brain. Your brain. The part of you that you are using to read and understand this sentence. The part of you that you use to perceive, understand, and interact with the world outside. The part of you that is truly you. Surely that is enough of a subject for the many, many books written about the brain. Yet I hope to do something different—we are often told how complex our brains are, but I want to convince you that it is, in fact, simpler than you might have thought. Not that there is anything wrong with simplicity—elegant simplicity can be very powerful.

Even scientific study of the brain is simple. Much of this book will be about how we have studied the brain over the centuries—or rather how the brain has studied itself. There are actually rather few ways to study that thing in our head, and those ways offer different rewards. First, we will find that there is a huge wealth of things we already know pretty much for sure about the brain (its geography, or structure). Second, there are things that we should probably accept we will never ever know for sure about the brain (its archaeology, or evolutionary history). And finally, there are many things that we are finding out now, or that we can see we have a good chance of finding out at some point in the future (its engineering, or function). When you discover what is known, knowable, and unknowable, you will truly understand the brain humming away in your skull.

But enough for now. It is time to make Rose's dinner. She is sitting here opposite me as I write this in our garden on a sunny summer afternoon. Or rather she was when I started writing. I was tapping away far too long and she wandered off to do something altogether more interesting. She is busy now, across the lawn, perceiving, thinking, and maybe even naming.

I

A GRAND TOUR OF TERRA INCOGNITA

The spinal cord itinerary:
Sailing from the horse's tail to the
mysterious obex

Well, what is this wonderful thing, and for what purpose has it
been made by a Nature who does nothing in vain?

—Galen, *On the Usefulness of Parts of the Body*

I

SKULL MARROW

First Thoughts about the Mind

Our first known reference to the brain has reached us by a tortuous route—inscribed on papyrus, sold surreptitiously by a temple on the Nile, bequeathed to a daughter and given up for translation early in the last century.

To stand in the center of the great step pyramid is to travel as far back as civilization can take you. Less famous than its later descendants at Giza, the step pyramid is from a different Egypt, when a society first tried to construct a neverending empire to dominate the world. This was the experimental stage of Egypt, before thousands of years of cultural conservatism made ancient Egypt the great changeless kingdom of our imagination. The step pyramid is far from town at dusty Saqqara, where it is easy to be alone with your thoughts, surrounded by these early experiments in monumental architecture.

There are beautiful stone courtyards, unfinished pyramids, wonky pyramids, and subterranean mausoleums full of entombed sacred bulls. I was deeply impressed by this rambling necropolis. Among this jumble is two hundred feet of gnarled, tumbledown grandeur. The step pyramid was the first time that an Egyptian architect experimented with piling up several slab-like mastaba tombs, each smaller than the last, to create a pyramidal shape. Although not as geometrically regular as later pyramids, in its own lumpen way the step pyramid is recognizably pointing upward; its creators clearly had their minds on matters celestial. These were truly ancient and obscure times, but a strained chain of archaeological links tells us that five thousand years ago this pyramid was built for a pharaoh named Djoser by a genius named Imhotep.

An even more fragmentary line of evidence leads us to wonder if Imhotep really spent all his time designing stone afterlife starships for his master. The mysterious papyrus containing the first known appearance of the word "brain" has also been ascribed to him. The Edwin Smith surgical papyrus was probably written around four thousand years ago, one thousand years after the death of Djoser and Imhotep. Yet it is thought to be a copy of an original text from their time, albeit a partial and modified copy. The fragmentary papyrus first appears in modern history in the late nineteenth century when it was sold in Luxor by one Mustapha Aga to the eponymous Edwin Smith, an American Egyptologist.

The papyrus was written in hieratic, the more cursive, everyday form of the neatly formal hieroglyphics we are used to seeing on monuments. Although ancient Egyptian writing had already been de-ciphered, translating hieratic is still extremely demanding. But Smith was Egyptologist enough to realize that the papyrus seemed to re-late to the subject of medicine—a rare enough thing in itself. The four-thousand-year-old papyrus with its five-thousand-year-old med-ical knowledge remained in his collection until he died in the early years of the twentieth century, when it was inherited by his daughter Leonora. In one of those frustrating delays in intellectual history, she kept the document for a further fourteen years until she donated it to the New-York Historical Society. Then the orientalist James Henry Breasted spent a decade working on the papyrus before finally pub-lishing a translation in 1930.

Reading Breasted's translation is to be swept back to an alien and violent world. The writing is pragmatic and concise, sometimes al-most terse, which rather adds to the feeling of strangeness that over-whelms the modern reader. This is practical text written by a practical man for coping with practical problems. The fact that these problems speak of almost unspeakable violence does not deter the author from his course. Clearly these were times when life was cheap. The papyrus is a series of medical case studies—forty-seven complete and one truncated—a guide to treating different types of physical injury. They include "a smash in his temple," "bulging tumours in his breast," and "a gaping wound in his shoulder," among other gems, although we do not know why the author had such a supply of horrible injuries. It has been suggested that Imhotep was master-surgeon to the Pharaoh's army and that these are the men who managed to limp back from the sandy battlefield. Alternatively, and perhaps even more worryingly, these men may be construction workers mutilated by accidents on some ancient building site. Maybe the text is describing what happens when large pieces of step pyramid fall on you.

There is one word that crops up several times in the papyrus, and it describes an especially vulnerable part of the body. It is thought that the word translates as something like "skull-offal," but we think that the author, maybe Imhotep himself, was referring to the brain beneath the skull. The word he used is:

Hieratic can be written left to right like modern English, or it may be written right to left like Arabic. This may seem unnecessarily confusing, but the ancient Egyptians had an obsession with symmetry. This bi-directional system of writing allowed them to frame their pictures with two mirror-image inscriptions, each meaning the same thing. Fortuitously, many hieroglyphic (and hieratic) symbols are actually little pictures of animals, and these were the clue to the direction of the text—the animals' noses point helpfully in the direction in which the writing is meant to be read. Things could be worse—some of the world's languages (ancient Hittite and Rongorongo, for example) change their direction of writing in each successive line—a form of writing called boustrophedonic: "as the oxen plow the field."

Anyway, the hieratic word here reads from right to left and contains four glyphs—"vulture," "reed," "folded cloth," and a final explanatory suffix meaning "little." Thus the individual symbols are not themselves related to the concept of "skull-offal," but signify sounds that together make up a word pronounced "ais"—just as the five letters in the English word "brain" do not have anything inherently cerebral about them. Tellingly, the hieratic word for "brain" is a derived word—made up from other concepts. The brain is described as the soft stuff in the skull, not of itself worth an original word. "Heart," "hand," and "eye" had unique words in their own right, but the brain did not—perhaps a sign that the ancient Egyptians did not think much of it.

The author of this papyrus has a matter-of-fact approach to his job. He examines, or "measures," each case and briskly allocates it to one of three groups: those that should get better if treated, those who might get better if treated, and those whom it would be best to leave untreated. This may seem harsh, but as his main forms of treatment are binding wounds with various combinations of linen, honey, grease, and meat, followed by asking patients to sit down for a while, perhaps his attitude was realistic. Strangely enough, I do recognize this approach in my own veterinary clinical work. Unfettered by the Hippocratic oath, I find that the main clinical decisions I make are deciding whether animals' ailments will be easy to treat, difficult to treat, or whether it would be kindest to end its suffering.

Our ancient doctor was very clear on two matters: wounds to the brain can be life-threatening, and the brain has unexpected effects on the rest of the body. He describes how a physician should feel head wounds and notice if he "finds something disturbing therein under his fingers, [and the patient] shudders exceedingly." The author knew that even serious bleeding was more likely to stop if a patient stayed still—notably allocating a patient who "discharges blood from both his nostrils [and] from both his ears" to the "may get better if treated" category.

The author also seemed able to use indirect evidence of brain function to make his decisions. A patient with an apparently less severe head wound but "with stiffness in his neck, so that he is unable to look at his two shoulders and his breast" was considered to be beyond hope. Another, whose "eye is askew . . . he walks shuffling with his sole, on the side of him having that injury which is in his skull" is also consigned to death, as is another man who cannot speak. All this has remarkable parallels in modern neurological examination. The unusual feature of brain examination is that clinicians are examining an organ that they cannot feel, hear, or see—at least until the advent of modern imaging techniques. All the information is gleaned indirectly from observing the brain's effects on the rest of the body. Neurologists today test exactly the same things our ancient doctor tested: can patients control the position of their eyes, and if not, which eye is affected; can patients understand speech and themselves speak; can they walk and manipulate objects normally, and if not is it because they are paralyzed or because they are unaware of the position of their limbs? I may only carry out neurological examinations on animals, but the way the papyrus's author investigates the goings on within the closed box of the skull seems remarkably familiar to me.

That said, at other times the writer certainly shows that he has an acquaintance with the tissue of the brain itself. Sometimes the closed box of the skull was itself smashed open and he describes tissues like "those corrugations which form in molten copper [and] something therein fluttering under thy fingers, like the weak place of an infant's crown before it becomes whole." This one sentence carries an astounding level of biological knowledge. First, it is clear from the comparison with the scum that forms on molten metal that he has actually seen the convoluted folds of the surface of the brain. The fluttering he feels presumably refers to the pulse, which might well be weak and irregular in a patient whose skull is split open. To link this fluttering to the normal pulse that can be felt at the gaps or "fontanelles" between the unfused bones of babies' skulls is a leap of impressive scientific sophistication. The editing hand who copied and annotated the papyrus a thousand years later adds further to our sense that the ancient Egyptians knew a great deal about the brain, stating that it is covered by protective membranes and has at its core a body of fluid, referring to a wound that penetrates "the membrane enveloping his brain, so that it breaks open the fluid in the interior of his head."

Completing this summary of third-dynasty neurology, the original author makes it clear that he knows that the spinal cord is the main link between brain and body. Just as he seems especially interested in the brain, he also displays what he knows about injuries to the spine. Once again a great deal of understanding is evident in just one sentence as he describes an injury that has "caused one vertebra to crush into the next one [so that he is] unconscious of his two arms and his two legs because of it." He realizes that if the bony column of the back collapses, then the limbs are often disconnected from their prime mover, the brain. In another case of spinal injury, he gives a prescient indication of the complexity of how the brain controls both sexual arousal and continence, describing an untreatable patient in which the "the phallus is erect and urine drips from his member." Once again there are parallels with modern neurology—today our major concerns following spinal injury remain the preservation of limb movement, continence, and sexual activity.

All this is very impressive—a man thinking in a very modern way in an ancient time—but there is a problem with the Edwin Smith surgical papyrus. The problem is that it is simply too practical. Its author obviously knew a great deal about the brain and spinal cord, but nowhere in surviving ancient Egyptian texts does anyone attempt to draw any general conclusions from all this knowledge. Nowhere in the papyrus does the eminent doctor change the course of medical history by simply wondering, "Isn't it remarkable that it seems to be the brain that senses the world and controls the body?" This is immensely frustrating. All the evidence is there to support a little bit of philosophizing about the function of the brain, but there is something holding the author back. Maybe his culture did not encourage natural philosophy, the generation of scientific theories from evidence of the world about us, or maybe the good doctor had more pressing matters to deal with.

So in some ways the papyrus was really a wasted opportunity. Not for another three thousand years would a great thinker reconsider the role of the brain. As far as we can tell, the Egyptians largely neglected the organ. Herodotus is the source of the popular story of how the Egyptians, when they were embalming a body, removed the brain through the nostrils and discarded it, but then again Herodotus did make some suspect statements for which we have no other sources. One thing we do know is that the brain was not afforded great care after death—certainly it was not one of the Pharaonic organs that was carefully sealed in its own canopic jar. The various funerary texts repeatedly refer to the importance of retaining the heart within the body for its owner to reach the next world. Rarely is the brain mentioned in any of the many Books of the Dead. Imhotep—if it really was he who wrote the papyrus—was so close to realizing that the brain is "us," but he never made that final philosophical step. For three millennia the brain was thrown away.

2

SERVANTS AND GUARDS
OF THE GREAT KING

The Classical Brain

An asklepion is usually a tranquil place today. Scattered around the Aegean, these temples to the physician-deity Asklepios are mostly dilapidated and overgrown and those I have visited seem for centuries to have been the domain of the scraggy goats that abound in this part of the world. A blue-skied, cool early morning in one of these places is about as refreshing as life gets, but these temples of healing were probably not always so idyllic.

As gods were wont to do, Asklepios could appear in animal form—it is he who hisses and writhes up the staff in the icon used by many medical organizations. Although gods are by definition immortal, rather confusingly he is said to have died when Zeus punished him for raising the dead. Clearly the live-to-dead transition is supposed to be a one-way journey. Perhaps in anticipation of this, Asklepios was claimed to be rather touchy about death, and it was not allowed to sully any of his temples—moribund incomers were excluded and dying inmates expelled. Still, they must have been fairly grim places in their day, when the sick, the halt, and the lame from the surrounding countryside converged on them. Yet it was in these places that modern western medicine started, and it retains its Greco-Roman flavor to this day.

There was once a man named Nicon, a well-to-do architect who lived in Pergamon on the western coast of what is now Turkey in the middle of the second century. One night Asklepios came to him in a dream and demanded that he encourage his son Claudius to study at the town's asklepion. Dreams were clearly an acceptable form of career advice in those days, so this is exactly what Nicon did. Claudius was immediately gripped by medicine, especially its theoretical basis in animal structure and function. Yet the boy was certainly no passive shrinking violet in thrall to his father. A driven, ambitious individual who was later to show a tendency for aggressive self-promotion, he was to become the most famous and influential physician in history, and his thinking was to dominate medicine for well over a millennium. Considering his strident manner, it is perhaps strange that he came to be known as "the gentle one," or "Galen."

Although what we know about Galen is dominated by his contribution to our understanding of how animals and people are put together and function, Galen made a career as an adept and successful clinician. It is difficult to say exactly when Galen ceased to be a learner and started to be a healer, as he probably did both activities throughout his adult life. His long medical education was as diverse as can be imagined. Financially independent following the death of his father, he embarked on almost a decade of what would now be called residencies in the great centers of medical learning of the eastern Mediterranean, practicing in Smyrna, Corinth, and Alexandria and ending with five-year stints in Athens and Pergamon itself. In the unlikely event that you wished to arrange an asklepion-based vacation, you could still visit many of these sites today.

This was not simply the classical equivalent of an over-extended backpacking trip by an overfunded student. The young Galen was constantly working, observing, and thinking. Again and again in his later written works, he refers to things he saw and ideas he gleaned on his travels. Galen had many medical interests and was at the cutting edge of almost all biology, but he was particularly interested in what drove animals to live and function. He built on an existing idea that "body spirits" pervade us and animate us, but instead of abstract theorizing he built his theories on observations of the inner structure of people and animals, both dead and alive. Not surprisingly, Galen's obsession with the motive forces behind life soon forced him to tackle the question of the roles of the heart and brain. As we will see, Galen more than any other told us what the brain does—that it is, in fact, where our selves reside.

The history of ancient Greek attitudes to the brain was rather stacked against Galen's ideas. The first sign that the brain was not an appealing concept to the Greek mind is their word for the organ itself: ἐγκερσλος. This transliterates to *enkephalos,* although the neuter form *enkephalon* is more often used by scientists. Like the "skull-offal" of the Egyptians, the Greek word is a derived one, but bland rather than dismissive in this case, meaning "in the head." Once again the brain is seen as some sort of padding for the inside of the skull rather than something of interest in its own right.

This is not to say that thinkers before Galen were entirely unaware of the importance of the brain. Homer included gory descriptions of characters dying from horrendous brain injuries in his epics, but this may reflect simple observation rather than any attempt at an explanation of the role of the brain. After all, anyone living in violent times would probably know that wounds to the head were especially likely to prove fatal. But at least one early natural philosopher, Alcmaeon of Croton, did make very specific links between the brain and perception, probably because he studied its connection to the eyes, the optic tract (often called the optic nerve). We know that he was not alone in his thinking, as others also speculated that the organ played a role in emotions and thought.

Aristotle was the chief proponent of a very different theory that placed the heart at the center of the body's processes—the cardiocentric theory. According to this, the heart was the prime motive force behind the body's actions and emotions. Of course, this idea persists in many English phrases today—most of us will suffer from a broken heart at some point in our life. The cardiocentric view is an appealing idea: the heart certainly seems very active. Only occasionally in the course of my veterinary work have I viewed the beating heart of a live animal, but it is certainly an impressive sight: an altogether vivacious and forceful organ. I have thankfully never seen the brain of a living animal, which would suggest that I had done something badly wrong, but I must admit that the brain really does not move nor make a sound. It is easy to see why Aristotle thought little of it.

Like Galen after him, Aristotle placed great importance in blood vessels, but came to very different conclusions as a result. First of all, he dismissed an earlier theory that the brain was the origin of all the body's blood vessels. This had been a fascinating theory, but it got the roles of the brain and heart completely the wrong way round. But Aristotle went too far when he said that "the brain in all animals is bloodless, devoid of veins, and naturally cold to the touch." All three statements were wrong, although some were more forgivable than others. Admittedly most of the large vessels supplying and draining the brain run in its surrounding membranes, and thus can be peeled away from the organ itself. However, there is a profuse blood supply to the organ, albeit distributed in tiny vessels. In fact we now know that the human brain receives as much as a fifth of the blood that the heart pumps around a resting body. So the brain can hardly be said to be "bloodless and cold," as anyone who had cut into the brain of a living animal would have known.

Aristotle did have other reasons for thinking that the heart was a more likely animator of the body—it is, for one thing, in a conveniently central position. He also reported that some "lower" animals do not have a brain, and so it cannot be essential for life—although it must be said that some animals do not have a heart either. In addition, he correctly pointed out that the heart forms very early in embryonic development, as befits a prime mover. However, we now know that the central nervous system (brain and spinal cord) starts to form even earlier, although as we will see later in this book it is probably the last to complete its formation. Yet to a scientist working without a microscope, the embryonic heart stands out for the same reason that it does in adults—it is vigorously pumping bright red stuff around while the nervous system is still just an indistinct pale strand along the embryo's back.

Aristotle put forward a charming idea, that the brain is actually present to cool the heart—rather like a radiator cooling the engine of a car. Certainly the brain has a corrugated surface, as does any good radiator. He repeatedly states that the heart is the "hottest" organ in the body and that the brain is the "coolest and moistest," and that these characteristics of the brain are most marked in humans, whose hearts allegedly churn out the most heat. Of course, with our modern understanding of physics we know that an organ which is actively cooling the body would instead feel relatively warm—a dog's tongue on a hot day, for example. But Aristotle not only relegates the brain from the position of "prime mover," he even states that it is the complete opposite of his chosen organ, the heart.

Yet even Aristotle could not escape the embarrassing fact that the sense organs of the head—the eye, ear, and nose—are clearly connected to the brain. He does mention this several times, for example wondering why he cannot find the connection between the nose and brain in fish. Yet he conveniently mentions the issue in contexts where he can avoid speculating about what these connections with the sense organs mean for the function of the brain. The grudging implication is that the brain acts as a conduit by which sensory information is fed into the spirits circulating around the body. But he believes this role to be supplementary to the brain's main job: radiating heat.

By the time Galen reached Rome, he was in his early thirties, and he was already unhappy with Aristotle's cardiocentric ideas. Still a very Greek thinker, he was to spend the rest of his life working at the center of the world's greatest empire. Although he continued to write in Greek, he mentioned his pleasure that Latin has its own distinct, original word for his beloved brain: cerebrum. At a remarkably young age for such a post, Galen was appointed imperial physician to Marcus Aurelius. Now the world's pre-eminent doctor, Galen knew he could challenge the authority of the father of natural philosophy. In his usual frank and confident tone, Galen laid into Aristotle's theories in his *On the Usefulness of Parts of the Body*.

First of all, he described himself as a true Aristotelian, committed to developing his ideas based on the evidence of his senses. After this brief homage to the master, he immediately turns on him and chides him for what he sees as his "amazing" lapse in his observations of the heart. Galen is adamant that the brain is not a heat sink for the heart—he counters that whether it is felt through skull fractures in humans or carefully exposed in a live animal in one of his own nightmarish experiments, the brain is indisputably warm, and indeed is damaged when chilled by the air. He also argues that the brain is too far from the heart, and too sequestered from it to act as a radiator:

> But even if [Nature] had been so negligent as to place the brain
> far away and attach the senses to it when there was no need of

doing so, she would certainly by no means have walled up the brain and heart in two such strong, safe enclosures.

To today's ears, this really does sound like Galen is coming to the right conclusion for the wrong reasons. Were the brain's role to shed heat generated by the heart, not only would it feel warm, but it would also be eminently sensible to place it at some distance from the chest—otherwise it might uselessly radiate heat straight back into the heart. This point about the futility of a radiator being enclosed in a bony box is a good one, however.

Galen's greatest contribution to modern biology is probably in his detailed anatomical descriptions—some of which are remarkably similar to those in today's anatomy textbooks. Maybe this is because his anatomy did not have to be distorted to fit into his theory of animals' spirits. Yet in Rome it was not permitted to dissect normal human bodies, and he had to be content merely to observe the aftermath of horrific injuries. Instead, most of his knowledge came from dissecting animals. He dissected not only the readily available corpses of domestic animals, but also apes. These were purchased especially for the purpose of dissection, because Galen realized that apes resemble humans more than other animals.

Whichever animal he dissected led him to the same conclusion about the brain: it has a very ordered internal structure. In fact, this order is suspiciously similar among all the mammals he studied, including his glimpses of the human brain. He realized that the basic arrangement of the nervous system is simple. Beneath its protective membrane, the brain is a soft but solid organ with interconnecting, fluid-filled central cavities. The brain stem is really just an elaborated upward continuation of the tubular spinal cord, but much of the mass of the brain is made up of a variety of extra lobes and protuberances growing out of that brain-stem tube. Galen also described how each of these protuberances had its own distinctive appearance, structure, and connections, and emphasized that the closer he looked, the more levels of new detail he saw.

We will use this idea of the structure of the brain as variations on a simple theme when we later start our tour of the organ, but for now what is important is the conclusion that Galen drew from this scheme. He simply could not reconcile the idea that the brain is a crude radiator of heat with his own observation that it has the most heterogeneous and complex structure of any organ in the body. Why have all these different shapes, textures, connections, and colors in an organ with one simple function? To Galen, this simply did not make sense. Galen was convinced that the brain was involved in sensation, perception, motivation, and action, which he saw as linked processes that must be coordinated inside a single organ. And taking exactly the same approach as modern neurobiologists, he focused initially on the first of these. There is something about the senses that makes them the easiest way to approach the brain. Harking back to Alcmaeon, he emphasizes that the proximity and connection of the senses to the brain must be highly significant:

> Does not a nerve of considerable size . . . enter each ear? Does not a portion of the brain much larger than that proceeding to the ears come to each side of the nose? Do not one soft nerve and one hard one come to each eye, the former inserted into its root and the latter into the muscles moving it? . . . Hence all the instruments of the senses . . . communicate with the brain.

He then went on to analyze the functions of almost all the myriad nerves connecting the brain to the various structures in the head, and claimed he had also found a route by which taste sensations may reach the brain from the tongue. To Galen, the status of the brain as the central point to which all the senses project was extremely important. He described the sense organs clustering around the brain as being like "the servants and guards of a Great King." The brain was finally, he believed, to receive the respect it deserved. His simile also openly suggests a very interesting idea—that the retinue of sense organs are mostly in the head because the brain is there. Or is it the other way round? As I will discuss in a later chapter, it is now thought that much of the evolution of our bodies has been based on the need to cluster our sense organs and brain in one place.

Of course, a brain does not only sense, and we now think of sensation as just one footing of an arching bridge of activities that passes through perception, interpretation, motivation, and planning to touch the ground once again at action. Galen realized that the middle of this bridge was inherently difficult to study. How could he study the role of the brain in motivation, for example, with nothing more than a restrained animal and a knife? Instead he looked at the far end of the bridge, at actions and movements, which seemed altogether more amenable to study.

Galen was probably inspired to study movement as a result of witnessing some catastrophic surgical mistakes, although he assures us that they occurred at the hands of other surgeons. In a rather frightening passage, he describes how if a surgeon slips while applying pressure to the skull and a metal implement is accidentally thrust into the center of a living human brain, it instantly renders the patient immobile. Yet to Galen, accident was no substitute for experiment, so in a series of experiments on live animals, often performed in front of large public audiences, Galen began to demonstrate the role of the nervous system in movement.

Here the modern reader may start to feel uneasy. Galen had to perform experiments on live animals, and of course this was long before the invention of anesthetics. Although he was quick to consider the effects of his work on the animals themselves, he does seem rather removed from those effects. He discouraged experiments on live apes, but this was mainly because he thought that the facial expressions of apes in agony were too disturbing for a philosopher to watch, rather than because of any feelings the apes may have had on the matter. This may seem strange to us, but modern attitudes to animal experiments are similarly partial—in the United Kingdom, for example, it is more difficult to obtain an experimental license to work on primates, horses, dogs, and cats than to obtain a license to work on less human-like or cuddly animals. Also, working on the brain presents particular problems to the experimenter, even today. Most other organs can be studied in anesthetized animals, but anesthesia profoundly suppresses and alters brain activity—an anesthetized animal is obviously insensate and paralyzed to some extent. The possibility of anesthesia was not open to Galen, and we must simply accept that our current understanding of the brain, which has served us so well in the compassionate treatment of human and animal disease, is rooted in a harrowing series of second-century experiments on immobilized farmyard animals.

Galen experimented widely on nerves and gradually developed the idea that nerves can either return information to the brain from the organs of sight, sound, smell, taste, or touch, or transmit impulses to the muscles to make them contract. We now use the terms "sensory" and "motor" nerves to distinguish these two types. Galen took some time to reach this conclusion, probably because most nerves in the body carry both sensory and motor impulses.

Certain demonstrations clearly stuck in the public mind, and Galen recounts these with alacrity. In particular, he identified the two nerves (one on the left and one on the right) that supply the larynx, or voice box. For reasons to do with the embryonic development of the head and neck, these follow an exceptionally tortuous route. They exit the base of the brain and pass all the way down the neck into the chest. Here they execute a U-turn around some large blood vessels and course back up the neck to reach the muscles of the larynx. Thus, they were excellent targets for Galen because they were long, vulnerable, and responsible for obvious effects—the voice. Galen showed that when an animal was bound and turned over and its neck incised, it started to scream. This screaming continued until both laryngeal nerves were located, but stopped at the exact moment when both nerves had been cut.

Shocking as this procedure may sound, its implications were clear. The brain is connected to the rest of the body by many thin white cords, and the effect of these cords on the body is to animate it. Bloodsoaked but unbowed, Galen had established that the two ends of the sensation-perception-interpretation-motivation-planning-action arc are rooted in the brain. Also, he cannot have been unaware that the rest of the arc was probably located there too. He spent most of his time treating human patients, and he realized that brain injuries often result in disorientation, confusion, passivity, and unpredictability. Through the work of one man, the brain was elevated from something akin to a radiator to the central organ of the body—where all the action is. Where the self is, in fact.

Aristotle's cardiocentric theory is itself a good example of how ideas can take hold largely because some authoritative figure proposes them. Whereas the brain was held in little esteem in early Egypt and Greece, thinkers from other civilizations already had the impression that the self was located very much inside the head. In fact, many cultures, such as the Incas, employed trepanation, drilling holes into the skulls of living subjects, and we believe that at least some of these must have been attempts to release the evil spirits thought responsible for mental illness. In fact, many doctors in the classical Mediterranean world used trepanation (including Galen), but interestingly the brain's great champion himself recommended that its use be restricted to the relief of pressure inside the skull. Maybe he had seen with his own eyes how the procedure did not relieve mental illnesses, and it is to his credit that he did not attempt to use trepanation as a spurious justification for his ideas.

For all his achievements, Galen's insights into the brain can be seen as a triumph of hope over actual evidence. While his accumulated wealth of anatomical knowledge convinced him that the brain was the site of the whole sensation-to-action process, he could only provide evidence of the beginning and end of this process. Galen relied on his own optimism that the evidence of the inner workings of the brain would emerge. And that hopeful confidence was to keep brain scientists going for the next seventeen centuries. In that time, many thinkers added further layers of detail to our understanding of the structure of the brain, each contributing to an overwhelming sense that something elegant and wonderful must be going on in the organ. For what is more wonderful than thought itself? Yet even after the invention of the microscope, there was simply no way to study the actual processes going on inside the brain. For seventeen hundred years of European history, the brain was all structure and no function—a rococo ensemble of buff jelly with nothing to do.

3

THE BRAIN AS GEOGRAPHY

Maps of the Mind

We talk of process and states, and leave their nature undecided. Sometimes perhaps we will know more about them—we think. But that is just what commits us to a particular way of looking at the matter.

—Ludwig Wittgenstein, *Philosophical Investigations*

Form and function. Substance and activity. Galen was showing us the best way to think about the brain, although not in the way he had hoped. His writing is filled with descriptions of how the different spirits—natural, vital, and animal—are generated in the body and travel around it. These spirits formed the core of how he thought the body keeps itself alive and is moved, but we now know that they were illusory. We now know there are forces that drive the body—electrical and chemical interactions in every cell—but they do not correspond to Galen's spirits. In the classical world there was not yet enough accumulated understanding of physics and chemistry to comprehend that the forces that act within living bodies are exactly the same as those acting throughout the rest of the universe. Electricity and chemistry are forces enough, so there is no need for a system of special spirits to animate us.

Instead, what Galen showed us was that concentrating on the matter of the brain itself—its structure, heterogeneity, and order—is the best way for us to start thinking about the brain. If we start by trying to work out how the brain can undergo perception, consciousness, and other mental processes, then we will soon be overwhelmed by its apparent complexity. How can a jelly-like mass do all these miraculous things? Even today when we have many technological ways to investigate brains, starting our search by studying its function is a staggering task. So before the twentieth century such an approach would have been entirely futile.

This is why, for seventeen centuries, scientists tried the alternative approach. They realized there was something that they could truly know about the brain—its structure. The ancient Greeks had shown that the different parts of the brain are intriguingly variable in texture, color, and arrangement. There are soft bits, hard bits, folded bits, dangling and protruding bits, fibrous bits, granular bits, pale bits, red, black, and even blue bits. Galen's hunch that all this internal structural complexity must be important was what kept brain scientists going all those years. And the closer they examined this structure, the more they learned, yet the more it seemed there was to be learned. Closer and closer inspection revealed finer and finer detail, and never did the anatomists reach a level of detail at which the different parts of the brain blurred together and looked alike. Even after the invention of the microscope, the different regions of the organ still retained their peculiar local characteristics.

And all this time the anatomy of the brain was all we had—and the only barriers to an ever-increasing knowledge of its structure were the limitations of existing investigative techniques. There seemed no theoretical limit to the depth in which humans could understand the brain's anatomy, and yet for most of the written history of science this stood in direct contrast to our laughably rudimentary ideas about how the organ actually worked. Today it is difficult to appreciate how anatomy was for so long not only our best avenue to the brain—it was our only avenue.

As I mentioned in the prologue, I hope to convince you that the structure of the brain is still the best way to understand it. Of course, I would say that—I am an anatomist—but I hope to show it really is true for all sorts of reasons. First, I will show you that the anatomy of the brain is surprisingly easy to understand if we take it step by step. Once you appreciate that, you can start to see the organ's internal pathways forming in your mind's eye—the wiring diagram of the brain. Again and again in the brain we find bundles of fibers linking up different regions, and in many cases we know exactly what happens when those bundles and regions are damaged. Looked at this way, you will soon be able to visualize the brain as an array of tangible units acting in concert to do all the clever things that a brain can do— and a reasonably manageable array at that. Some people like to impress you with the complexity of the brain, but I will show you that, much of the time, it is really far simpler than you might have thought and often does straightforward things.

The other advantage of our impending assault on the brain is that when we investigate the structure of the brain we immediately get a huge amount of extra information for free. This is because there is not just one type of brain in the world but tens of thousands. The noggin of a guppy, a toucan, and yourself do not all look exactly the same, but as we will see again and again, their basic organization is similar, although it is overlaid by modifications that allow them to pursue their guppyish, toucanesque, and human lives. The story of the vertebrate brain is very much one of variations on a theme, and those variations are immensely informative—we have learned a great deal from the alterations to brain structure that allow animals to swim, fly, burrow, see colors, vocalize, and dispense with senses when they no longer need them. And we will see that, as well as the variations, the theme itself is important too—how can such a diverse array of creatures function with the same sort of brain?

But I do not want to tell this story simply because I think that anatomy is the only accessible way that someone can explain the brain to you. After all, just making a subject clear does not necessarily make people want to read a whole book about it. Instead, I wanted to tell you the story because it takes place in a deliciously weird world. Although you might think it unlikely, for reasons that will soon become clear, our understanding of how the brain is put together has an endearing charm to it—a charm that few other fields of intellectual endeavor can match. Of course the history of brain science has a spooky, self-referential edge to it, as the seething brain continually strives to pick apart its own inner workings. Yet this spookiness is more than balanced by the touchingly simple and homespun ways in which we approached the task. And the reassuring story of that search is etched forever into our map of the brain.

Most of the charm of neuroscience, as we should perhaps now call it, derives from the skewed way in which it progressed. As we have already seen, for most of the time we have been studying the brain, all we knew about was its structure, with almost nothing coherent known about its function. This has had one strange consequence for our view of the brain—almost all of its parts were named before we had the slightest idea what they did. Thus, whenever we speak of parts of the brain—and we will speak of many of them in this book—we usually have to use words that have nothing at all to do with what we are trying to say about those parts. For me the fact that the internal parts of the brain are labeled with names entirely dislocated from their function is what makes the brain so wonderful, and it has led to two delightful results.

The first of these is that the brain is endowed with a set of names that puts all other fields of science to shame. To the people who named the parts of the brain, those parts were essentially mysterious, so they could effectively name them as they wished. Also, there was no recommended system of naming and so each namer used a different approach to the named. Never again in human experience would we have the opportunity to catalog such an interesting object so precisely in such complete ignorance. And never again would we end up with such an emotive lexicon of names—rather appropriately for the organ that is, after all, the seat of emotion. Some names are simple ("the almond") and some haunting ("the marriage chamber"). Some sound exotic ("the tract of Goll") and others just inexplicable ("brain sand"). Some seem functional ("the bridge") whereas others in the same neck of the woods seem overly so ("the nucleus motorius dissipatus formationis reticularis of Riley"). Some of the best are wistful ("the field of Forel"), mysterious ("area 23 of Hippo"), or just plain defeatist ("substantia innominata" or "unnamed stuff"). Again and again in this book we will encounter places with dramatic, questing names, adding up to a veritable geography of a mysterious organ.

And like the hopeful scattering of untraveled alien places around the edges of a medieval map—terra incognita—it is the romance of the accumulated cataloging of the unknown that inspires.

As well as all those names, there has been another strange result of exploring the geography of the brain before we understood what it actually does. Does the fact that our map of the brain is full of names unrelated to function affect how we think about it? The place names in our geography of the brain may be varied and colorful, but do they sometimes get in the way? We humans are verbal creatures, so it seems likely that the names we use can influence the thoughts we have—a rather unpleasant possibility for modern scientists who strive to study the brain without preconceptions. Philosophers of language have long wondered about how tagging certain entities with particular names skews our thought processes about those entities— see, for example, the quotation by Wittgenstein at the beginning of this chapter. This problem is especially severe in the study of the brain. We really do have a dual view of the brain: one an objective view of substance, texture, and function, and another a verbal folklore of names acquired by quaint historical accident over the centuries. Here more than anywhere, scientists must constantly juggle two systems in their heads, allowing the names to explain but not alter their work on the organ itself—they must practice a form of doublethink.

This is not the first time that naming things has threatened to interfere with their objective study. I have already briefly referred to the wonderfully named and often fictitious people and places that ancient mapmakers used to fill the gaps in their knowledge at the periphery of their world. Like the names used in the brain they were the products of ignorance, but unlike brain names many of them were based on vague travelers' tales or were acts of deliberate fiction. And yet the giant unipeds, the mandrakes, the basilisks, and all the other monsters that swarm around these *mappae mundi* are what made these documents so compelling to the adventurers willing to risk all to explore terra incognita. The names given by one generation of brain scientists inspired investigations by succeeding generations, intrigued by the marvelous names given to structures with unknown functions.

An even more ancient parallel with the naming of the brain lies in the constellations. Just like the parts of the brain, the stars were named and their arrangement cataloged long before anyone understood their composition or origins. There is clearly a very strong drive to group the stars together into constellations because it has been done again and again in different cultures around the world—usually with very different results. Yet unlike the brain, this repeated cataloging was based on a complete misconception. We now know that the clustering of stars into constellations is illusory, an attempt by the human brain to create order in an array of random dots that are constantly on the move, albeit extremely slowly. But here again modern scientists still use the constellations to guide them around the sky and to describe positions in the heavens. And intriguingly, just as in the brain, they give this practice a veneer of respectability by using Latin—perhaps "Ursa Minor" sounds more scientific than "Little Bear."

This problem of false names has not ended with the stars, Earth, and the brain. There is another, more contemporary field of intellectual activity in which names are assigned long before the named entity is fully understood—the naming of genes, or the little molecular instructions that tell every cell in our body what to do. The folklore of gene naming has been concocted in two phases. Last century, most genes were discovered because of something they did, and thus were often given names that now seem eminently sensible. Of course, this function-based naming is the opposite of what took place with the brain, but this still does not mean that gene names are entirely logical. For one thing, many genes so named have since been found to have other and often more important roles, and for this reason their old names now make little sense. For example, some genes have several different jobs at different stages of embryonic development. I once did an undergraduate project on a developmental gene called *hairy,* which is an important player in establishing the segmental structure of insects, but was originally named for causing a developmental defect you can probably guess at. In addition, some gene names were assigned at a time when scientists were free to give names with a predictably short cultural shelf life that now seem cringingly dated. The developmental gene *sonic hedgehog* is perhaps the best example of this. Nothing dates like fashion.

The second age of gene-naming folklore has come in the twenty-first century as the Human Genome Project has completed the mapping of the human genome. Surprisingly, this methodical cataloging of the structure and contents of the human genetic instruction manual has much more to do with anatomy than function. Although a remarkable achievement, it has given us an exhaustive list of human genes, most of which have no known role. The similarities to all those centuries of hopeful brain anatomy are obvious. We now have names or at least codes for all (23,000 or so) of these genes, even if we cannot explain what they do. At present many are cataloged because their structure seems similar to genes we already understand, but such assumptions can be dangerous. At least the architects of the project are entirely honest that their genome sequence, their anatomy of our DNA, is no more than that—a tentative map with which to explore.

All that said, despite being overlaid with rich history, language, and legend, it is still the anatomy of the brain that will give us the clearest way of approaching how that exceptional organ works. As in any journey, the language and history should be seen not as hindrances, but as local color. An unhealthy obsession with how the brain works seems to me like running before one can walk. After all, the structure of the brain has so many advantages—it is eminently tangible, easily explained, intriguingly diverse, awash with folklore, and to a great extent already known. We will worry about the knowability or unknowability of other aspects of the brain in later chapters, but for now one thing is clear—unlike other fields of neuroscience, the anatomy of the brain is relatively accepted. It is as close to a fact as anything in science can be. It is a far more stable platform on which to start than by worrying about consciousness, intelligence, emotion, self, or soul—all entities or processes that we cannot even define properly, and that we rarely use in our daily lives. I will consider these aspects of the brain in later chapters, but I will continually bring them into the realm of the real by considering where, if anywhere, they are actually located.

The idea of being able to localize specific mental functions to particular parts of the brain has a long and illustrious history, even though we really knew very little about it until recently. For example, Galen's theories of spirits fueled a long-running and ultimately futile search for the seat of the soul. Some later thinkers thought that the soul was located in the meshwork of blood vessels Galen had observed at the base of the brain—the *rete mirabile* or "marvelous net." They had simply overinterpreted the information Galen had left them. Galen had dissected many different animals, but human material was hard to come by. Humans differ from most domestic animals in that they do not have a *rete mirabile,* and so by this theory they, along with dogs and horses, should be devoid of the soul so evident in cats, pigs, sheep, and oxen.

One of the most famous attempts to localize brain functions was the nineteenth-century pseudoscience of phrenology. According to its theories, complex human traits such as conscientiousness and acquisitiveness were localized within and effected from defined little patches of brain. Phrenology was immensely popular on both sides of the Atlantic, and a multitude of phrenological systems appeared. One of the reasons for the popularity of phrenology was that it claimed to give insights into differences among individuals. If a person had particular personality traits or habits, then the parts of their brain associated with these would be correspondingly enlarged: a philanthropist would have a prominent Organ of Benevolence, a pickpocket an enlarged Region of Acquisitiveness. And as the skull was said to develop according to the arrangement of the brain beneath, these traits and characteristics would manifest as readily identifiable lumps and bumps on that person's head.

Although we now realize that the claims of phrenology were, perhaps disappointingly, entirely spurious, its descendants are still with us today. As we tour the geography of the brain in this book, we will see that in many cases discrete functions do indeed seem to be linked to discrete locations in our heads. Some of these will be rather unsurprising—such as the clusters of sound-processing cells near where the nerves from the ears enter the brain. Other examples are more striking—stimulation or damage to small parts of the brain can evoke or destroy memories of specific musical passages, affect abilities to conjugate verbs, or even cause experiences of déjà vu.

Localization of brain function in this way must be viewed with caution. Admittedly we have learned a great deal from electrodes and injuries, and it is probably fair to assume that if electrically stimulating a region of the brain stimulates a process, or if damaging that region prevents it, then that region may be involved in that activity. However, it does not mean that this process is entirely localized in that place. We cannot assume that all brain processes can be pinned down to a neat little chunk of tissue, and as we will see, there is considerable evidence that some brain functions may be mediated by large circuits of cells dispersed around the brain. One important example is that there does not seem to be a region that acts as the common route for all brain activity—we have no equivalent of the central processing unit of computers.

Throughout this book I will discuss this issue of how functions may be restricted to regions of the brain, or how some of them are spread more diffusely throughout its substance. Maybe this is quite a good indicator of how evolution has provided us with the brain we have. Obviously an arrangement in which each function was tied to a particular region would be conceptually simple, but maybe there are advantages to be gained by spreading and intermingling activities across the organ. For example, might a widely dispersed process be less prone to irreparable damage by small, localized brain injuries?

This is an area in which computers might help us. There is much written about how our attempts at artificial intelligence have informed our study of the brain. However, I am more interested in the nitty-gritty of how we went about constructing these information-processing machines to help the pre-existing information-processing machine in our heads. One thing is very clear to anyone who has used a computer for an extended period of time—computers are extremely vulnerable to damage. My skull has had things dropped on it, my brain has been infused with alcohol and exposed to viruses. Yet it still functions to some extent. And my brain has never broken down irreparably after sucking cat fluff into its cooling fan. The structural elements of a computer must be almost perfect for it to work at all. Clever software can counteract a small amount of damage, but even then only in certain components.

In complete contrast, the brain is designed to withstand damage. It may never even form optimally in the first place, but it is certainly destined for a lifetime of damage and degeneration. Every time my son whacks his head on the ground I try and forget the huge numbers of brain cells that are claimed to die during such an event. And even in my late thirties, I am very much aware of how I am already trying to shore up my waning mental acuity, agility, and energy with the senile comforts of experience, perspective, and concentration. Loss is built into the brain's very operation—for example, there seems to be a process by which memories are permanently discarded, and we will see later what a helpful process this loss is. In short, the brain is meant to be in a constant state of coping with failure. Although it may seem sad, this is one of its greatest achievements.

So the brain is a well-charted organ, named and misnamed. We think we know what it does, but we are not yet certain. It does some things that we cannot even define properly. But much of the time it seems to do comprehensible things—taking in sensations, processing them, and then making the body do sensible things in response. We have not yet seen where any of these functions are located in the brain, nor even if we can expect them to be neatly localized. The brain is very good at coping with damage and decay, and even seems to thrive on it sometimes. Oh, and like the rest of the body, it develops from almost nothing.

4

A RIVER RUNS THROUGH IT

The Development of a Brain

There is one feature above all others that helps to make the brain seem simpler, and we have known about it since the time of the step pyramid. The Edwin Smith surgical papyrus briefly mentions a body of fluid at the center of the brain, and almost everything written since then has made mention of the organ's inherently hollow nature.

A cut in almost any direction through almost any part of the brain or spinal cord will expose at least one little channel or chamber of clear, colorless, odorless fluid. More detailed study will show that all these cavities are in fact interconnected into a continuous system of internal watercourses. Early biologists were fascinated by this "ventricular system" because it was as mysterious as it is elaborate. Galen had even considered it an important repository for his spirits. All backboned animals have this fluid, but it was not immediately obvious where the fluid is made, or where it goes, or what it is for. Most pockets of liquid in the body have an obvious destination toward which they are pumped or disgorged. Not so the fluid at the core of the brain. It is embryonic development, not plumbing, that tells us where it comes from.

In the nineteenth century, anatomists changed the focus of their work from studying the finished adult form to investigating how that form is established in developing embryos. By 1827 both the human sperm and egg had been identified, and it was now clear that the complex architecture of a human baby is formed by a dramatic elaboration of a tiny and presumably rather amorphous fertilized egg. At the time it must have seemed astounding that such a tiny nugget of life could somehow organize and pattern itself into a living, breathing baby with all its thousands of interlaced, interacting body tissues, and to be honest, it still does. Anatomy had told us how complicated a body is—now it was time for embryology to tell us how it gets that way.

As early embryologists worked on ever more immature unborn babies and animals, the initial signs were not encouraging. Progressively earlier human fetuses just seemed to be, well, progressively smaller. A baby halfway through pregnancy appeared much like a fully formed baby, except that it is, of course, very small. In fact, this is true for much of pregnancy—nine weeks after fertilization a fetus looks like a miniature baby, just over an inch long. It seemed as if most of the arranging of body parts in a child is completed within these first nine weeks (when the baby is called an embryo), after which the baby simply grows (and is called a fetus).

Amazed that the fabrication of the human form could take place in only nine weeks, embryologists started the meticulous process of picking apart exactly what was happening in the first trimester of pregnancy. In the first few days after the sperm hustles its way into the egg, the resulting hybrid cell starts to split, first into two, then four, then eight cells. Within five days the few hundred cells thus generated start to specialize into different types. First of all the solid, globular embryo cavitates into a hollow ball of cells containing a blob of cells within. The outer hollow ball then spends much of the second week negotiating board and lodgings with the mother's womb while the inner cell blob flattens into a disk (Figure 4.1) and takes its first tentative steps at turning into something recognizably babylike. Already, there are only seven weeks to go until those magical nine weeks are up, and there is still no hint of a brain, hollow or otherwise.

The tiny disk hangs suspended between two small, fluid-filled cavities. The cavity below the disk is called the yolk sac and is very important in feeding embryonic birds, reptiles, and fish, but in humans it almost disappears by the time of birth. Above the disk lies another

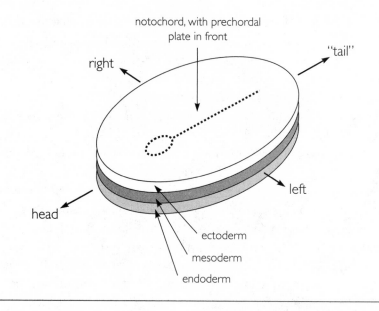

Figure 4.1. The developing human embryo, showing the three-layered human disk-embryo from above. It already has a front-back and left-right orientation, mainly apparent at this stage from the notochord and prechordal plate lying deep in the mesoderm layer. The ectoderm layer will form most of the brain and spinal cord.

cavity that is much more important to us—the amnion. This cavity will grow to surround the developing baby, and its fluid acts as a kind of shock absorber to protect it from impacts. This is why unborn babies often escape relatively unscathed when their mothers suffer a serious accident. Humans are quite unusual among mammals in that the amniotic sac is the only body of fluid surrounding the baby by the time it is born—it contains the "waters" that "break" so spectacularly before birth. Most other mammals have a second fluid sac—the allantois, or "sausage"—but it is still the amnion that enshrouds the embryo most closely, and many infants are born with the wall of this sac still stuck to their fur or wool. This is probably how the amnion got its name—it means "lamb."

Although not an obviously promising start, the embryonic disk soon starts to do some cellular origami that makes it look a little more like an animal in miniature. Actually, origami is putting it rather too simply. Cells can execute all sorts of contortions in the embryo fabri-

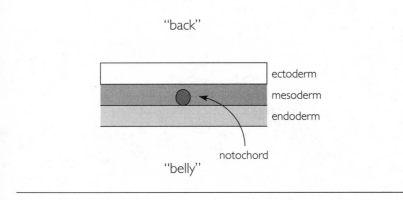

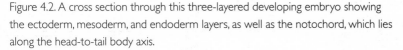

Figure 4.2. A cross section through this three-layered developing embryo showing the ectoderm, mesoderm, and endoderm layers, as well as the notochord, which lies along the head-to-tail body axis.

cation process. Not only can they fold, but different parts of the embryo can grow at different rates, distorting the entire structure. Also, clusters of cells can detach from their neighbors and migrate to an entirely different position in the embryo.

Initially, the disk has two layers, rather like a sandwich with no filling. Soon a groove—the primitive streak—appears on its top surface, and cells from the top layer flood through this groove and push the cells in the lower layer out of the way. This replacement of the lower layer is not the end of the cellular exodus from the top layer, however, and a second wave of emigrants now floods through to fill the space between the upper and the new lower layer. We now have, about sixteen days after fertilization, a three-layered disk—a filling has appeared in the sandwich (Figure 4.2).

This three-layered disk is a very important stage in our formation as these layers will play distinctive roles in our later life. As it happens, this is a very ancient system—most animals form their bodies from three slabs of cells in this way. Jellyfish, hydra, corals, and sponges are just about the only animals that do not. The cells left in the top layer are called the ectoderm, or "outer skin," and they are important to us because roughly half of them will form the brain and spinal cord. The other half will form the outer layer of the skin as well as important parts of many sense organs. The lower layer is the

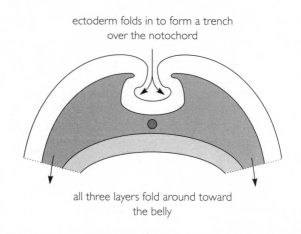

ectoderm folds in to form a trench
over the notochord

all three layers fold around toward
the belly

Figure 4.3. A cross section of the embryonic disk as its edges start to curl downward and inward. At the top of the embryo a trench has formed above the notochord— this will become the central nervous system.

endoderm, or "inner skin," and it is destined to make the lining of the gut, lungs, and bladder. The filling of our strange embryonic sandwich is the mesoderm, or "middle skin," and it has the daunting task of building most of everything else in the body.

To make something animal-shaped, this unprepossessing three-layered disk must curl up into a tube, like a rather stale ham sandwich. The left and right edges of the disk start to curl downward, toward where the belly of the embryo will eventually lie (Figure 4.3). A similar curling also takes place at the front and back of the embryo, and all this curling continues until the edges of the disk actually touch each other underneath the embryo. This folding has now left the embryo with its outer surface lined by ectoderm and has sealed off an endoderm-lined gut inside—this is why these layers are called the outer and inner skins.

While all this gut folding is going on, something exciting is taking place on the back of the embryo. Embedded deep inside the center of the embryo is a cord of mesoderm running from head to tail: the notochord. Along with some structures at the head end, this starts to induce changes in the ectoderm overlying it. This is what embryonic cells are good at—releasing chemicals to make nearby cells do things they otherwise would not. This release of controller-chemicals is responsible for much of the patterning of embryos, although these chemicals cannot spread very far. This ability to act only at close range is probably why most body formation goes on when embryos are very small—so that the distances are small enough for these all-important chemicals to seep across.

Although you may not be aware that you have a notochord, this structure has a long and noble history. The major subdivision of the animal kingdom in which we and all other vertebrates lie is called the chordates, or animals that have a notochord. It is the structure that defines us. The notochord probably first evolved as a gel-filled tube running the length of our ancient boneless aquatic ancestors. It could not be shortened, but it flexed easily from side to side and thus made an excellent framework to which muscles could attach in our sinuous swimming forebears. Ever since that time, the notochord has been with us, dividing us into left and right halves—showing in which direction we are going. What better structure to organize the development of a nervous system running from head to tail? The role of the notochord in making us bilaterally symmetrical creatures with a head and a tail is most clearly demonstrated by the fact that the closest living relatives of the chordates, our closest relatives without a notochord, are starfish and urchins. You may well wonder where your proudly historic notochord is now—once it has laid out our embryonic body plan, we treat our notochord with disdain. In humans it ends up as the jelly-like part of the disks between our vertebrae—the bit that pops out when we "slip a disk."

Under the influence of the chemicals from the trusty notochord, the ectoderm along the middle of the embryo's back starts to thicken until a large slab called the neural plate is visible along its length. This soon starts to crease, creating a shallow groove all the way along the back (see Figure 4.3). This groove is continually bathed in the same amniotic fluid that soaks the rest of the top surface of the embryo. The groove gradually deepens to form a veritable trench as more and more of the neural plate is folded into its depths, a process called neurulation. As the infolding of the neural plate becomes ever more dramatic, the two edges of the trench are bent toward each other and eventually touch. They then stick together, sealing up the skin of

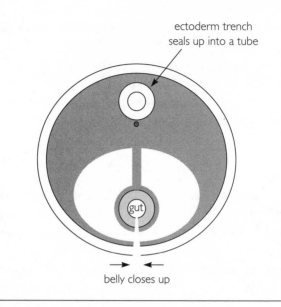

ectoderm trench
seals up into a tube

gut

belly closes up

Figure 4.4. A later cross section of the embryonic disk. The downward and inward folding continues until the belly closes up. The trench at the top of the embryo has now sealed over, leaving a buried ectoderm tube—the central nervous system.

the back and isolating the infolded neural plate as a tube running along the length of the embryo (Figure 4.4). At this stage, the tube still opens into the amniotic cavity at its front and back ends—the neuropores—but these seal up by the twenty-eighth day, completely isolating the neural tube from the outside world.

Although a sealed tube buried in the back may seem a very simple structure, the neural tube is actually the key to understanding the configuration of the adult brain and spinal cord. Indeed, it is no exaggeration to say that the central nervous system you now have in your head and back is simply an elaboration of that simple embryonic structure. The ectoderm walls of the tube will thicken and form almost all the cells with which you perceive, think, and move. And the space within the tube, once isolated from the amniotic fluid, will become the ventricles of the brain—the fluid-filled central cavity described by the Egyptians and Galen. And that is all there is to it. Admittedly we will have to flesh out the details a little, but if you can hold in your head the image of that tube running just under the surface of the embryonic back, then you are well on the way to understanding how a brain is put together.

In emphasizing the importance of the tubular nature of the nervous system, I have deliberately avoided a few complications to this story. First of all, although most of the tube forms this way, for reasons that are not clear, small stretches of the tube form in a different way. Folding is not the only way to make a tubular structure, and an embryo can also make a neural tube by forming a thickened cord, which is then excavated along its core. This is exactly how you made the lowest region of your spinal cord, the part that plugs into the small of your back. Different groups of vertebrates use this cord-excavation method for varied and often extensive stretches of their nervous system, but we do not really know why. It may seem unnecessarily complicated to use two different methods to achieve the same result, but that is exactly what happens. After all, there is no rule that says that nature has to be simple.

I also may have given you the impression that one part of the neural tube forms pretty much like any other—that a simple uniform groove deepens and pinches off into a uniform linear tube. In fact, there is no time at which the neural tube is completely uniform along its length. This is because different parts of the tube have different fates—the lower part will become the spinal cord whereas the upper parts will become the progressively higher regions of the brain. These different destinies mean that some parts of the neural tube will have to grow faster than others—and it seems that they cannot wait until after closure of the tube to start becoming different from each other. Even at the flat-slab neural plate stage, the upper, brainy end of the plate is already noticeably larger and wider than the lower, cordy part. It already seems to know what it will form, even at this early stage when the primordial nervous system is still connected to the surrounding skin. And even within the early brain-slab itself, little sub-bulges are already visible, corresponding to subsections of the future brain.

A final oversimplification I have made is to imply that the neural groove seals up into a tube in one fell swoop—that one day it is open and the next it is closed. In fact, in all vertebrates who form their nervous system by folding, there seems to be a general rule that different parts of the tube seal up at different times. It is almost as if embryos have a series of zippers closing up their backs and heads—each zipper with its own starting place, time, and direction of "zipping." This hotchpotch of neural closure has some strange results. For example, your back and the top of your head closed up before the region between them did, and you were left for some time with a gaping hole at the back of your neck.

Learning about embryos can be a process entailing two steps forward and one step back. Formation of a complicated little being from a fertilized egg is a true miracle, but it is hardly surprising that such a carefully coordinated process can sometimes go wrong. In fact, the more one learns about the intricacies of embryonic development, the more one wonders how it ever works at all. The whole process is remarkably robust, but for every story of elegant embryonic construction there is a tale of tragic failure. The folding of the neural tube is no exception. Abnormalities of the neurulation process are among the best known of all birth defects—after heart defects, they are probably the most frequent. And as I mentioned in the foreword to this book, these neural tube defects are also one of the main reasons why we ultrasound scan developing babies.

In neural tube defects, the tube simply fails to seal up properly. One of the zippers fails to zip. Or rather one of two zippers fails to zip, as two different regions of the nervous system are most commonly affected. If the region of tube corresponding to the top of the head does not fuse, the result is anencephaly, or "no brain," and if the defect is in the lower back, the condition is called spina bifida, or "cleft spine." Perplexingly, these two most error-prone zippers are not meant to be the latest nor the earliest to close—there just seems to be something inherently unreliable about them.

Although less well known, anencephaly is probably the more common of the two abnormalities, occurring in maybe 0.1 percent of all live births, and also causing perhaps twice as many stillbirths and miscarriages. Its effects are usually fairly catastrophic. Because most of the brain part of the tube does not infold, anencephalic babies' brains usually amount to little more than a flat plate of malformed tissue on the back and top of their head. The vault of the skull does not form as there is no brain for it to surmount, and so most of the top of the head looks as if it is missing. If they are born alive, these infants rarely survive more than a few hours, although some may survive considerably longer—even years—with extensive medical support. In many cases, anencephalic pregnancies are terminated after the obvious signs show up on ultrasound, but attitudes vary as to how to manage live-born babies with anencephaly. There can be little doubt that around the world most are allowed to fade away, yet many people vociferously advocate medical support of these children, even though the condition is incurable and it is unclear to what extent they suffer.

Spina bifida is much more variable, ranging from almost undetectable to extremely severe. The mild and often undiagnosed forms are most common, and this is one reason why it is difficult to say how frequently the condition occurs—perhaps 0.03 percent of births is a reasonable estimate. In the most mild form, spina bifida occulta ("hidden"), the lower spinal cord does manage to seal over, but it may still be abnormal. In this variant of the condition, the most obvious evidence may be a tuft of hair over the lower spine or a failure of some of the protective vertebrae to completely form around the spinal cord—a tongue-twister of a condition called rachischisis, literally a "schism in the stem (spine)." Children with spina bifida occulta may be partially paralyzed or incontinent, but then again they may not. They may even improve with a little reconstructive surgery. Worse problems arise when the contents of the vertebrae—the spinal cord or its surrounding meningeal membranes—protrude out of the defect in the bony canal. Depending on what protrudes, this can be called myelocoele ("marrow-cavity") or just meningocoele. Even within myelocoeles there is considerable variation—the spinal cord may be essentially normal, but covered only by a thin membrane that gives it little protection. Alternatively, the cord may have failed entirely to seal into a tube, and is only present as a deep cleft or flat plate on the baby's back—a frozen relic of embryonic development.

These problems aside, let us return to the story of our developing embryo. By the time it has sealed up, then, the neural tube is already well on the way toward specializing into the different parts of an adult brain. By this stage, at the thirtieth day in humans, its top end has already swollen into three little hollow bulges, the forebrain, midbrain, and hindbrain, the last of which connects to the spinal cord

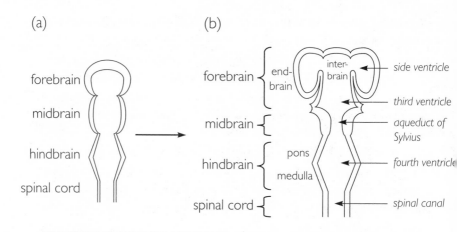

Figure 4.5. The developing human brain: Viewed from the front, the brain is seen to divide into three (a) and then five (b) swellings, each containing their own fluid-filled ventricle.

(Figure 4.5a). These three regions also have Greek scientific names—the prosencephalon, mesencephalon, and rhombencephalon. The first two are simply translations of the English name, and the last just refers to the diamond or rhombus shape of the hindbrain. As in so many things in embryonic development, this three-bulge brain is a reflection of our evolutionary past—all vertebrates, from dogfish to man, start with a three-bulge brain. In later chapters we will see why this is so.

But why settle for three bulges when you can have more? Almost as soon as the brain has reached this stage, it starts to progress further into a five-bulge structure (Figure 4.5b). Two lobes start to grow out of the left and right sides of the forebrain bulge, drawing a part of the fluid-filled space inside the brain with them. These two lobes are the endbrains (telencephalon), and as we will see, they have a big future ahead of them. When you look at an adult human brain, almost all you can see is endbrains. The bit of forebrain left in the middle, soon to be overshadowed by its neighbors, is called the interbrain (diencephalon). Farther down, the midbrain doesn't really change much, but the hindbrain is now starting to divide into the upper pons and lower medulla. These two latter regions are also called the metencephalon or "beyond-brain" and the myelencephalon or "marrow-brain," but we will not use these terms much. There is no clear boundary between the pons and medulla—they merge imperceptibly into each another, but we will later see why they are considered to be different.

All this time the fluid-filled core of the brain tube is becoming more and more complex in shape. By the five-bulge stage most of the fluid is held in four ventricles: a "lateral" or side ventricle in each of the endbrains, a third ventricle in the interbrain, and a fourth ventricle in the hindbrain. The side ventricles connect with the third through Monro's holes, also called foramina, and the third and the fourth connect via the classical-sounding aqueduct of Sylvius through the midbrain. In my diagram it looks as if most of the brain fluid will pass from the fourth ventricle into the canal running along the spinal cord, yet in the finished brain, most of it actually leaks out of the brain altogether through little holes in the fourth ventricle— Magendie and Luschka's holes. As you can see already, neuroanatomists like to leave their mark on the things they discover.

We had always assumed that this proliferation of lobes and bulges was a result of some parts of the neural tube simply growing faster than others. After all, as we have seen, the different bulges are prefigured by widenings in specific regions of the old neural plate, which presumably result from increased growth in those areas. Once the tube is sealed, however, we now think that the various brain bulges may form by a previously unexpected mechanism. Rather strangely, in many animals the canal at the center of the neural tube becomes blocked somewhere around the junction between the brain and the spinal cord. This occurs immediately before the period when most bulge formation takes place. This could be explained away as a coincidence were it not for the discovery that this tube-blocking happens in the embryos of a wide variety of vertebrates. Could it be that blocking the outlet of the ventricular system of the brain is essential for the formation of its different parts? It now seems likely that this temporary blockage to the drainage of the brain leads to an increase in the pressure in the fluid in the ventricles, and that it is this pressure that drives the expansion of the early brain. In other words, the early brain may not so much grow outward as be inflated from within.

These brain bulges do not inflate and sprout in a vacuum, however. They are developing in the restricted space available inside the embryonic head and, as it happens, at this early, bulgy stage the brain is growing faster than the head that imprisons it. Because of this it must fold and twist just to fit inside the head—rather like one of those acts in which a contortionist packs himself into a suitcase. The brain can manage this feat of packaging, but not without becoming a rather grotesquely squashed and gnarled thing, and some of this squashing persists into adult life. For an organ designed to interact with the outside world, the brain is remarkably confined.

These folding contortions are most obvious when viewed from the side (Figure 4.6a, b, c). A dramatic ninety-degree bend forms in the midbrain as the forebrain starts to outgrow the head. Once the endbrains start to sprout from the forebrain, they cannot grow forever sideways, but are compressed back onto the rest of the brain. In a mature brain, the endbrains almost enwrap all the other parts. Each endbrain grows not only sideways, but also forward and backward, and the backward growth is so rapid that the hindmost parts have to bend forward again to find enough space (the dotted arrow in Figure 4.6c). This gives each endbrain an appearance not unlike the shape of a boxing glove, with an extra lobe curling around from the back like the thumb of the glove. This thumb, called the temporal lobe, is especially well developed in people and does all sorts of fascinating things. Yet the reason it bulges out is simply that embryos' brains are cramped.

Although it is not growing quite as fast, the hindbrain is still squashed. The pons and medulla of the hindbrain flex forward and form a kink between them. As this kinking occurs, the back surface of the hindbrain almost tears open, as if from the strain of the kink. This can be a difficult process to explain, and I will revisit it later. It is, as with many scientific phenomena, best demonstrated with a banana. Hold a banana in both hands—one hand at each end—and try to bend it so that it curves the "wrong" way. Obviously, the banana will resist this rebending and will split open along its usually convex surface. While the usually convex side of the banana gapes open and drops dollops of mushed banana on your feet, the usually concave surface will remain intact, just like the front side of the hindbrain.

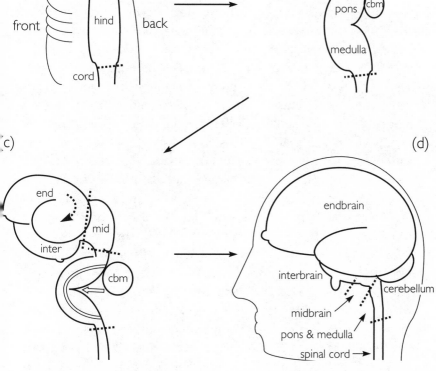

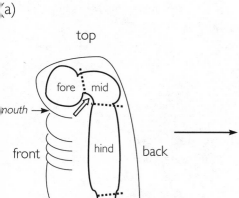

Figure 4.6. The developing human brain: *(a & b)* Viewed from the left side (as if the embryo is facing to the left of the page), the same bulges may be seen, along with the kinks that form in the midbrain and hindbrain regions. *(c & d)* Later, the endbrains and cerebellum start to form and come to dominate the brain. The other regions are just visible below the end-brains and are collectively known as the brain stem. The endbrains, especially, are starting to outgrow the other regions and their hindmost parts have bent around to create a boxing-glove shape.

Although the back side of the developing hindbrain does gape like our mutilated banana, mercifully it does not extrude mushed brain through the nape of the embryo's neck. Instead, a thin layer of membrane persists over the back of the hindbrain, retaining the fluid in the fourth ventricle. This membrane is so thin that it is transparent, and if a few other things are held out of the way, you can peer through it into the diamond-shaped fourth ventricle.

So the fourth ventricle of the hindbrain achieves its diamond-shaped form as the back surface of the neural tube almost splits open, leaving a thin roof membrane holding the ventricular fluid in. All around this membrane are thickened rims of brain tissue rather suggestively called rhombic lips, and it is from these that the last major subdivision of the brain forms. A globular structure soon starts to bulge out of the lips at the top end of the ventricle. It grows extremely rapidly and by the time a baby is born this globe is, apart from the endbrains, the largest part of the brain. Because of its size and the convolutions that form on its surface, this sixth bulge is called the cerebellum, or "little brain." The nerve cells within the cerebellum proliferate even faster than the cerebellum itself swells, so it ends up being more densely packed with cells than anywhere else in the brain. Remarkably, it may even contain more nerve cells in total than all the rest of the brain, although cells can be difficult to count accurately. The astounding growth of the cerebellum makes it very vulnerable when something retards an embryo's development—it is expanding at the absolute limit, and it feels any restriction acutely. In many species, viral infections can stifle development of the cerebellum, reducing its ability to carry out its postnatal role of controlling movement. For example, lambs infected with a certain virus before birth may be born with a syndrome called "hairy shaker"—they have coarse wool and tremble. Even after birth the cerebellum remains a viral target—my nephew suffered from cerebellar inflammation after a bout of chicken pox, but more of that story later.

The embryo now has all the basic components of its brain in place—five bulges and a sixth, the cerebellum perched on the back. Only two steps remain to convert the brain into its final configuration by the ninth week of embryonic life (see Figures 4.6d and 4.7). First, the kink in the hindbrain simply unkinks, presumably as the head

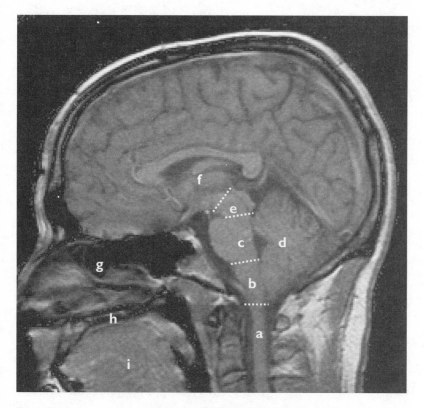

Figure 4.7. A magnetic resonance image (MRI) of my own brain, taken by my long-time friend and now neuropsychiatrist Dr. Nick Medford. The image is of a "slice" taken through my head from front to back, with my nose on the left. My spinal cord (a) can be seen widening into my medulla (b) and pons (c), behind which is the small dark triangle of my fourth ventricle. My cerebellum (d) appears in this view to be "floating" behind my pons, and some of the fine corrugations in its surface can be seen. My pons leads into my midbrain (e), within which the thin, dark line of my aqueduct of Sylvius is visible. Two "hillocks" are visible on the back of the midbrain. My midbrain goes through a near-ninety-degree bend before the aqueduct widens into the third ventricle within my rather heterogeneous interbrain (f). Above the interbrain lies my homogeneously pale corpus callosum, and the rest of the cranium is filled with the largest region, my cerebrum, in which the convolutions are obvious. Also visible are my nasal cavity (g), mouth (h), and tongue (i).

grows and frees up more space for it. Second, the endbrains grow immensely faster than all the rest of the brain until they are completely dominant. In the diagram you can see that the endbrains override all the other parts, even the cerebellum, and the interbrain, midbrain, and hindbrain just peek out from underneath. All backboned animals have endbrains, but they become especially dominant in mammals, and unusually so in humans. Intriguingly, the endbrains seem to be particularly large and corrugated in mammals that we like to think are intelligent, but there are exceptions to this rule. The huge size of the endbrains is probably the reason why we ended up calling the adult endbrains the cerebrum, which simply means "brain," as if no other parts were of any consequence. And to add insult to injury, we often dismissively lump the other parts of the brain—hindbrain, midbrain, and interbrain—together as the brain stem.

Later, when we dabble in consciousness, individuality, language, and all the things we like to think make humans special, there is something worth remembering about all the folding, swelling, and flexing of the developing brain: All animals with backbones go through almost exactly the same processes (Figure 4.8). The six-bulge brain is a common thread throughout vertebrate evolution—a remarkable concordance first discovered in the early nineteenth century by the very same man who discovered the human egg, Karl Ernst von Baer. Yes, we may exploit those six bulges in different ways, but we all have essentially the same layout. The differences between us and other backboned animals are just a matter of degree. Even the supposedly characteristic mammalian overgrowth of the endbrain has occurred time and time again in other groups—birds, some bony fish, some sharks—and sometimes almost to the same extent as in humans. The take-home message of the development of our brain is that there seems to be very little that is exceptional about the human brain at all.

There is one way in which the human brain is unusual, however, although it hardly seems likely to explain our apparent braininess. You will remember that the embryo forms a ninety-degree bend in its midbrain and forebrain, apparently so that it can fit into the restricted space inside the head. In humans, this kink remains into adulthood, but in most animals it unfolds during embryonic life. I spend most of my time teaching veterinary students, and so the retention of the

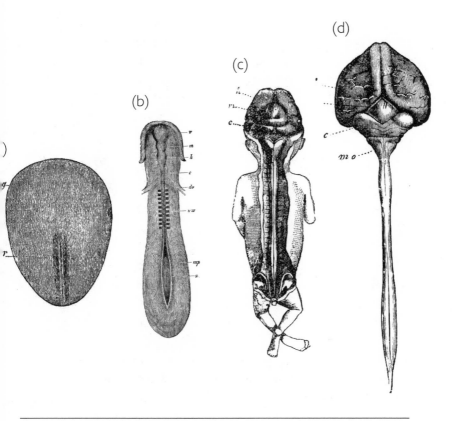

Figure 4.8. Four developing vertebrate nervous systems viewed from the back (all from a nineteenth-century embryology book). (a) Rabbit embryo at the disk stage. (b) Chick embryo, showing the long, deep neural trench in the lower back. Higher up the embryo, the tube has already closed, and the three bulges of the fore-, mid-, and hindbrain are already visible. (c) Human embryo at three months, dissected to show a very adultlike central nervous system. (d) Human nervous system at four months. The two large lobes of the endbrain, or cerebrum, are visible at the top. Below these, part of the midbrain is visible through a diamond-shaped gap. The boomerang shape below this is the cerebellum. Below the cerebellum, the cone-shaped medulla leads into the long spinal cord.

brain kink in people seems very confusing to me. Animals' adult brains are much more straightforward—they are laid out along a single nose-to-tail axis with no sudden bends. The reason people are different is literally so we can get our head straight. Unlike most animals we have an erect posture, and this means that our entire head has

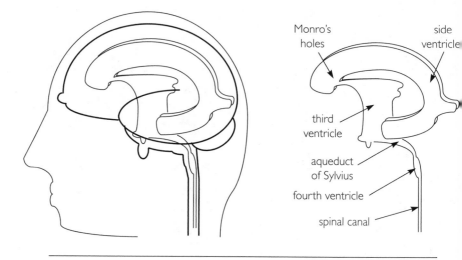

Figure 4.9. The final configuration of the ventricular system inside the brain *(left)* and also isolated from it *(right)*.

been bent around a ninety-degree corner. If you stand gazing into the distance, your eyes are looking along a line at a right angle to the spinal cord running down your back. Yet if your dog did the same, his gaze would be parallel to his spinal cord. This is why most animals have straight brains whereas we have bent ones. The bend relates to posture, not intelligence. Other vertically oriented animals share our brain kink—many other primates, some marsupials, and presumably seahorses, too—although admittedly the kink is not always in the same place. This is really the only aspect of early brain development that varies much among vertebrate species—in fact, amphibians and fish never have a kink in the first place, so I assume seahorses must form their own special kink.

Throughout these nine weeks of folding, swelling, and contortion, the brain is hollow. And from this time until it finally switches off and dies, the brain retains its original tubular conformation, and has a body of liquid within it. By the time the arrangement of the brain is established, the various twists and turns of its development have distorted the ventricles at its core into a complex shape (Figure 4.9). Most changed are the side ventricles, as the rapid growth and bending of each endbrain into its boxing-glove shape has stretched the ventricles within into elongated, recurved shapes, not unlike rams' horns. They still connect to the third ventricle via Monro's holes, but that ventricle is now a vertically oriented chamber—a high, narrow corridor almost completely separating the left and right sides of the interbrain. The third ventricle has a few nooks and crannies around its periphery, but we will worry about those later. It still connects to the top end of the fourth ventricle via the narrow aqueduct of Sylvius through the midbrain. And that fourth ventricle is still a flat, diamond-shaped lake covered by a thin, transparent membrane, emptying from its lower end into the narrow canal running down the spinal cord.

From the time when the fluid inside the ventricles pinches off from the amniotic fluid bathing the embryo's back, it has a new name: cerebrospinal fluid. Cerebrospinal fluid does not just sit there—it flows and is constantly refreshed and discarded. It is actively secreted into all four ventricles by delicate little pink fronds of blood vessel called choroid plexi. This name is derived from the way these plexi, or "plaits," often encircle the fluid-filled spaces of the ventricles. Rather delightfully, "choroid" means "to dance in a circle," as in "choreography," and we will see that this metaphor has appealed to anatomists more than once. The production of fluid by the plexi creates a pressure that actively forces the fluid through the brain from front to back, where it escapes. Most of it leaks out through those holes of Magendie and Luschka to flood the space under the meningeal coatings of the brain, whence it is drained. Not all of it leaks through the holes, however, and a little makes its way all the way down the spinal cord to leak out of its lower end.

The cerebrospinal fluid's recipe is carefully controlled. It probably carries chemical signals from the upper parts of the brain to the lower. As we will see, it is also a route by which parts of the brain may "sniff" the chemical state of the body. Because the fluid leaks out under the meninges, there is also a thin layer of fluid surrounding the brain. In fact, the brain actually floats in this shock-absorbing fluid—the organ has few direct connections with the skull around it, but instead bobs about in a fluid bath. Because of this, if the head makes a sudden jolt or twist the brain can move slightly within the skull. It is slightly cushioned, and this is a major protection against concussion. Finally, if the pressure inside the skull starts to increase for some reason, the cerebrospinal fluid can give a limited, temporary respite— tiny amounts of fluid can escape down or around the spinal cord where it exits the skull, and relieve the pressure.

We often hear about cerebrospinal fluid in a more negative context, when the internal plumbing of the brain fails. If the fluid is produced too fast, or more commonly if fluid cannot escape from the brain, the ventricles swell until they start to press on the surrounding brain tissue. This is hydrocephalus, literally "water-brain"—as you can see, most medical terms are so simple that they must be concealed behind translation into a classical language.

Hydrocephalus is often congenital, and babies born with the condition usually have an obstruction somewhere in their ventricular system. The most common site for these blockages is the bottleneck of the aqueduct of Sylvius. In this case, cerebrospinal fluid cannot pass from the third to the fourth ventricle. Less often, Magendie and Luschka's holes may not form, so fluid cannot escape from the fourth ventricle to seep under the meninges, as in "Dandy-Walker syndrome," a genetic disease causing malformations in organs all around the body. Congenital hydrocephalus varies in its severity, but in some cases the ballooning ventricles can almost entirely obliterate the cerebrum. The brain tissue is squashed so thinly around the clear fluid of the swollen ventricles that light can pass right through the baby's head. A major symptom of hydrocephalus in babies is that their head becomes larger as the pressure from the ventricles forces the unfused flat bones of the immature cranium apart. The progressive damage to the brain tissue leads to vomiting, fits, drowsiness, coma, and even death.

Fortunately, there is a surgical treatment that can relieve the pressure of cerebrospinal fluid—this is simply a plumbing problem, after all. Surgeons can insert a "shunt"—a tube that drains fluid from inside the brain and runs under the skin to a distant site where fluid can be safely dumped, usually in the abdomen. This seems like a clever solution, but these shunts block within ten years or so when debris accumulates in the tubes and valves. This can be a very dangerous time. It seems that a brain relieved of excess pressure for several years actually becomes less able to cope with that pressure, should it return. These children show many of the same signs as before, but if their pupil sizes, blood pressure, heart rate, or breathing become abnormal, then this is a very bad sign—we will see that these are evidence of damage to the brain stem. Hasty surgical reduction of pressure can even worsen the problem, as the brain stem may suddenly shift or even tear as the pressure plummets. For these reasons, neurosurgeons are constantly on the lookout for alternative ways of managing congenital hydrocephalus, and in the future simpler procedures may be used, such as punching a little drainage hole in the floor of the third ventricle.

Hydrocephalus does not only occur in babies—the ventricular system can be dammed by injury, blood clots, swellings, and tumors in people of any age. The effects are similar, except that the brain seems less able to cope with the squashing. The mature head cannot swell because the skull is now fused into a rigid case. All a semi-solid brain can do when it is put under pressure is exude out of the back of the skull, with predictably disastrous results. The best treatment for noncongenital hydrocephalus is usually a mixture of relieving the pressure and removing the obstruction, whatever it is. Occasionally, the injuries that cause the hydrocephalus can create their own curative drainage routes—for example, cerebrospinal fluid can ooze through skull fractures into the nose, from which it then drips, sometimes for years.

There is one last twist to the hollowness of the nervous system—it can become the lodgings of a bizarre visitor. We have seen that the brain forms by an infolding of the embryonic back. As a side-issue we also saw that the gut and body cavities form as foldings on the belly-side of an embryo. Folding seems an eminently sensible way to make a hollow tube, but very, very occasionally it can have some extremely unexpected effects—something can get accidentally trapped in that tube as it seals shut. You may wonder what exactly is drifting around in a pregnant womb waiting to be sealed inside a developing tubular organ, and I am afraid I have to tell you that the answer is a sibling.

We do not know how often one embryo gets drawn inside another, but it is probably a rare event. Certainly, the condition is very rarely diagnosed—we have no idea how many internal brothers or sisters are simply too small to be noticed. We assume that these siblings are in fact identical twins of their host, as nonidentical twins would probably be separated by too many membranes for entrapment to be possible. This strange fraternal internalization even has a suitably eerie name: *fetus in fetu*. In most cases the twin gets trapped in the abdomen, where it anchors and becomes parasitic, usually looking like a cyst, but in mind-bogglingly rare cases it is trapped inside the folding neural tube and persists within the ventricular system of the brain. And there can be more than one—as many as five fetuses have been reported inside a single brain.

There are, it must be said, disbelievers of the *fetus in fetu* phenomenon. Some of these internalized parasitic babies look so abnormal that some claim they were never separate individuals at all. Instead they may be teratomas—large but often benign tumors containing a variety of tissues, including skin, hair, and malformed teeth. However, some contain so many body parts that it is difficult to explain them away as tumors. Also, the tumor theory cannot account for why one baby should end up with five of these things in its brain, although admittedly the internalized twin explanation is not really much help either. One thing is clear: *fetus in fetu* is a reminder of just how remote and unfamiliar the world of the embryo can be. And there is something distinctly unnerving about the idea that some of us may be carrying a little passenger in our head.

5

LEONARDO'S BUTTERFLY

The Spinal Cord

It is time to embark on our journey through your central nervous system. We now know the basics of what the nervous system does and we also know quite a lot about how it forms, so we are ready to stride forth. The journey will fill most of the rest of this book, and will take us from the bottom of the spinal cord to the front of the brain. Yet, like all the best trips, it is the little detours into weird local backwaters, customs, and mythology that reward us most. As with so many things in life, it is not where we are going that is important. It is the manner in which we travel.

Our first stop is the spinal cord. Its simple, tubular shape reminds us of the early days when the neural tube folded up, as well as the era maybe 600 million years ago when our ancestors' brains were just tubes without a brain at the front. Yet we must not think of the cord as just a drab connecting lead between brain and body—as we will see, there is a great deal of activity here. Admittedly, humans are among the backboned animals for which the spinal cord probably does the least, but we are not typical. We all know that chickens have the dubious advantage over us that if they are decapitated, they can still run. More impressively, the sorts of people who decapitate and skin snakes have reported that the neck stump can "strike" at exactly the region of snake that is being skinned. It seems that spinal cords can even get angry.

The mature human spinal cord is a thick-walled tube with a narrow central canal down which flows tiny amounts of cerebrospinal fluid throughout childhood and adolescence, at least until it becomes blocked early in adult life. From the place where it dangles out of the foramen magnum, or "big hole," at the base of your skull, the cord is roughly seventeen inches long. At regular intervals, spinal nerves emerge from left and right sides of the cord and spread out to reach almost all parts of the body. There is a pair of spinal nerves corresponding to each of your bony vertebrae, and these nerves exit the spine through the gaps between each segment of your spine.

You probably will have noticed that your back is longer than seventeen inches, so this means that your spinal cord cannot reach all the way down it. It usually stops at the level of what is called your second lumbar vertebra, but if you want to know where that is, get a tape measure and measure the seventeen inches down from the base of your skull. This discrepancy between the spinal cord and the bony spine around it is simply due to the fact that the spine grows faster than the cord during embryonic life. You need not worry about the fact that the cord does not extend farther down your back, because the nerves destined to exit the vertebral column down there simply have to track all the way down to the appropriate vertebra before they emerge from the spine. Hence the cavity inside your lower back vertebrae contains no cord, but instead carries a neat little bundle of nerves diligently coursing downward to where they are supposed to escape from the bony spine. These nerves fan out slightly, creating what is poetically called the *cauda equina*, or "horse's tail."

The sluggish growth of the spinal cord, failing to keep pace with the vertebrae around it, is almost the opposite of what we saw in the embryonic brain, which almost outgrew its container. Yet this relative shortening of the cord is not just a human phenomenon. It appears that most vertebrate body plans follow it to some extent. There may even be good reason for the shortness of the cord—it may make animals more resilient to injury. One of the most important practical differences between the central nervous system and the nerves which flow in and out of it is that nerves in the periphery can usually heal and regrow, whereas the brain or spinal cord cannot. If you damage one of the large nerves running down your leg, then there is a distinct possibility that the nerve will grow back downward from the injury site, as long as it is not too disrupted. In fact, if neurologists know where the injury is, they can even make a pretty good prediction of when the nerve will reach its target, as nerves regrow at a quite reliable rate.

Yet as we all know, damaged spinal cords are not as resourceful. If the cord is severed, then it usually stays severed. The central nervous system is frustratingly obstinate in this failure to heal, a weakness clearly demonstrated by the fact that peripheral nerves have been shown to grow back toward the spinal cord, but then come to an abrupt halt when they attempt to grow into it. An immense amount of ongoing research has revealed that there may be something about the environment within the central nervous system which stops nerve cells growing once an animal reaches maturity. We do not know why we have evolved nonhealing spinal cords and brains in the first place—is an ability to heal not always a good thing? Perhaps nerve growth after maturity would be more trouble than it is worth—would our brains become clogged with exuberant tangles of overzealous nerve cells?

Is this why the spinal cord is shorter than the spine? Could it be that over the course of evolution we have shrunk our vulnerable, unhealing spinal cord as much as possible and compensated by lengthening some resilient, healable nerves instead? We cannot really prove theories like this, but it is true that the spinal cord is the most exposed part of the central nervous system, and that the cord is shortened in many vertebrate species. The most dramatic example of this is the sunfish. Apart from sharks, this creature is the largest fish in the world, measuring up to ten feet long and weighing up to one and a half tons. Despite their size the spinal cord of a sunfish can be less than an inch long, just poking out of the back of its skull—and the rest of the spinal column is filled with a profuse horse's tail of nerves. Why these leviathans should be so in need of such a stumpy cord is not clear—certainly it seems unlikely they are particularly prone to back injuries. Maybe the sunfish is telling us that cord shortening is not a safety feature after all. Anyway, the relative inability of the brain and spinal cord to heal is very real, and is something to which we shall return.

The width of the cord can be quite variable. The average human cord is roughly half an inch across, but its width varies along its length (see Figure 4.8d). At the bottom end it tapers to a stubby point called the conus medullaris, or "marrow's cone," as the nerves of the horse's tail fan out from it. Higher up are two swellings, or intumescences, in the regions that correspond to the lower neck and the small of the back. These intumescences are the parts of the cord whence large nerves travel to the arms and legs, and the swellings are there simply because more cells are needed to deal with the greater amounts of information going in and out of the cord in these regions. Almost all land vertebrates have intumescences, with the unsurprising exceptions of snakes and other limbless types. Birds have larger intumescences for their forelimbs—their wings—than for their "hindlimbs," with the exception of the large flightless emus, rheas, and ostriches, who quite reasonably have the opposite arrangement. If you liked to read books about dinosaurs in your youth, you may remember the oft-quoted unscientific statement that stegosaurus—the supposedly stupid herbivores with armored plates on their backs and spikes on their tails—had brains in their skulls the size of a walnut, but a larger second brain above their hips. Scientists now think this "second brain" observed in stegosaur fossils was a cavity in their vertebrae that contained a hindlimb intumescence ten times larger than their brain. Stegosaurs thus still had only one brain, and were probably not given to intellectual debate, but they must have been pretty slick at walking.

Like the rest of the central nervous system, the spinal cord is surrounded by the meninges. We mainly hear about these membranes when they get inflamed or infected in meningitis, but their principal role is to protect the spinal cord. There are three layers of meninges in mammals, birds, and reptiles, but fish and amphibians only have two. The innermost membrane was obviously named by someone of a religious bent—it is the pia mater, or "tender mother," perhaps because it envelops the central nervous system so closely and with such selfless care. The transparent pia is, in fact, so intimately stuck to the central nervous system that if you had not been told it was there, you would probably miss it. The next layer out is the more spooky-sounding arachnoid mater, or "cobweb mother"—a good de-

scription of a thin but clearly visible, lacy membrane. The space between the pia and arachnoid contains the cerebrospinal fluid that has
leaked from the ventricles, and it is from here that the fluid drains
into the blood vessels or can be sucked out by a needle in the procedure of lumbar puncture. The outermost membrane is the dura mater, or "hard mother," a reference to the thick, fibrous nature of this
meninx. The dura provides much of the physical protection conferred
by the meninges, but the space between it and the arachnoid is also
the site of the feared subdural hematoma—a blood blister that can
form after injuries and compress the underlying nervous system with
potentially fatal consequences.

There is one important difference between the meninges and the
central nervous system they enshroud—they can feel pain. You may
have heard that the brain itself is paradoxically insensitive to pain,
and indeed if you ignore its blood vessels, this common assertion is
probably true. Yet the meninges that cover it are extremely sensitive.
For example, much of the pain of a slipped disk comes from the
meninges, and none of it from the squashed cord itself. Also, many
headaches may be a result of pain in the meninges—although we
know surprisingly little about what causes headaches. All this does
have one very practical consequence for neurosurgeons—if they an-
esthetize the skin, bone, and meninges as they incise into a conscious
patient, they may then operate freely once they reach the brain or
cord. Because of this, perhaps chillingly, surgeons sometimes operate
on the central nervous system—the very fabric of what it is to be us—
while their patients are conscious. There are often very good reasons
why it is advantageous to have a wakeful patient during a brain operation, and we will see in Chapter 16 how such operations have been
invaluable in telling us about the brain's inner workings.

Back to the cord itself. Whatever processing may go on in the spinal cord of people, chickens, or snakes, one of its main functions is to
convey information to and from the rest of the body. This realization
dawned on different ancient cultures at different times. The most impressive exhibits in the entire British Museum are the Assyrian friezes
purloined from the North Palace of Assurbanipal in what is now Iraq.
Dating from 645 B.C., they are one of the most spectacular works of
art I have ever seen—huge stone plaques depicting dramatic scenes of

a royal lion hunt. But just one of the lions depicted is important to us here—her front end stands proud, roaring at her attackers, but her back has been pierced by arrows and her hind legs drag uselessly along the ground. To me, her pose looks exactly like a cat that has had its back broken by a car, and clearly shows that the Assyrians realized that the spine is the route by which animals control their limbs. Early Greek thinkers seem a little more confused. Initially they thought that the spinal cord was connected to the urinary and reproductive systems—some even claimed it was the source of semen. It was not until one of Galen's heroes, Hierophilus, asserted that the cord is an extension of the brain and carries signals between it and the body that Greek understanding of the cord could progress.

This two-way flow of information is very much built into the way the cord is wired up (Figure 5.1). At intervals along the cord, two nerve trunks are attached on the left side and two on the right. The back one on each side is mainly for receiving information from the body—it is called the posterior, or back, or even sensory root. In contrast, the front one on each side sends information out to the body, usually to make it move, or do something—this is the anterior, or front, or motor root. Although the exact number varies among species, the human spinal cord has thirty-one such sets of roots from nape to bottom, and these correspond to the thirty-or-so gaps between our vertebrae. There are two minor complications to this neat system: there are no roots corresponding to the lower parts of the human tail or coccyx, and also for some reason the highest and lowest segments are lacking back sensory roots.

Almost as soon as a front and back pair of spinal roots have left the cord, they unite into a spinal nerve trunk. This escapes from the bony spinal column through the gap between two vertebrae and then continues the tree theme by branching off, spreading information to and collecting information from a region of the body. The nerve supply of the body is therefore segmental, with each side of the body divided up into thirty or so neck-to-toe stripes, each supplied by a particular spinal nerve—although there is admittedly some overlap between the domains of adjacent nerves. This zebra-like pattern of nerve supply usually has no outwardly visible signs, but this is not always the case. Herpes viruses, such as chicken pox, have a nasty habit of crawling

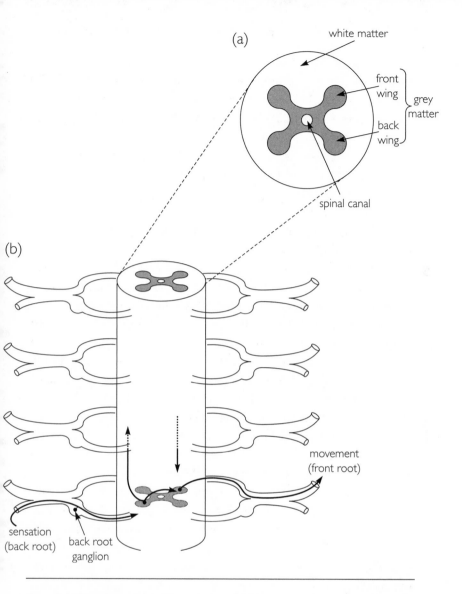

Figure 5.1. The arrangement of the spinal cord. (a) In cross section, an outer layer of white matter encloses an inner, butterfly-shaped core of grey matter, divided into front and back wings on each side. (b) This arrangement persists throughout the cord. Here, four segments of the spinal cord are viewed from the back. In each segment, a back and a front nerve root emerge on each side. These unite into a short, common trunk that then divides into branches. The back roots bring sensory information to the cord, and the front roots convey motor information from the cord. The main part of the motor nerve cells is in the cord itself, but the main part of the sensory nerves is in a little bulge on the back roots—the back root ganglion. Sensory information can pass up the cord to the brain, just as motor commands can pass down the cord from the brain. However, sensory inputs to the cord can also directly affect movements without any involvement of the brain (reflexes).

up sensory nerves and hiding out for years in the sensory nerve cells sitting in a little bulge in the back root—the ganglion. If you had chicken pox as a child, you will be delighted to hear that you have herpes viruses lounging in your back root ganglia, just waiting for a little bit of stress to rouse them to flood back down to reinfect the stripe of skin supplied by that spinal nerve. This is why shingles affects people in stripes.

The particular arrangement of the nervous system in humans, sensory at the back and motor at the front, runs very deep. In fact, just like the infolding neural tube of the embryo, it is a unifying theme that explains the layout of the entire central nervous system. When a spinal cord is cut in cross section, its internal structure is surprisingly evident to the naked eye (see Figure 5.1). Around the periphery is a pale, fatty region that we call white matter. Inside this is a central splodge of grey matter, with the tiny central canal at its center. The grey matter has a distinctive form—it is shaped rather like a grey butterfly contrasted against the white matter around it. Just as Leonardo da Vinci once drew an ideal human figure within his circle, neurobiologists have a butterfly circumscribed in theirs.

So why a butterfly? This configuration starts to make sense when you realize that grey and white matter are very different, and that this explains the internal arrangement of the cord. Individual nerve cells have a central soma or "body" in which most of the metabolic machinery of the cell is located, and from this body they often send out a single long, thin process or axon, which is what carries information to the other cells they are meant to influence. The grey matter of the cord is where the cell bodies are, whereas the white matter is where the bundles of axons are. The sensory nerves bringing information from the body enter the cord at the back, and so it makes sense that they connect to nerve cell bodies concentrated in the back wings of the grey matter of the cord. Similarly, the nerve cell bodies whose axons exit the cord in the front, motor nerve roots, are clustered in the front wings of the cord grey matter (see Figure 5.1).

So by simply looking at a cut section of cord, you can distinguish the areas containing the sensory and motor nerve cell bodies. This arrangement of sensory and motor cells is pivotal to understanding the brain and spinal cord, because it runs throughout these organs.

Encircling these butterfly wings are the white matter axons, either nipping in and out of the cord, or coursing up and down to the rest of the central nervous system. When we investigate these white matter axons, we find they too are regularly arranged into neat bundles, or tracts—some taking sensory information up to the brain and others bringing motor information back down. Unfortunately, modern neuroscience has tried to expunge some of their more interesting names—the tracts of Burdach, Clarke, Gowers, Lissauer, Flechsig, and my own favorite, the Tolkienesque Tract of Goll.

But as we have already seen, there is a great deal of evidence suggesting that the cord is not simply a passive conduit of information between brain and body. Your spinal cord constantly takes in sensory information, processes it, and uses it to control your actions. This initiation of actions without the involvement of conscious thought is what we call a reflex, and the cord is very good at reflexes. We often hear about reflexes—tapping our knee with a hammer, for example—as if they are some sort of freakish design fault that we can elicit for humorous effect. However, we are constantly using most of our reflexes, and for very good reasons. One of the easiest reflexes to explain is the withdrawal reflex—the simple protective reflex that makes us pull our hand away from noxious things, such as a hot stove. Not only does this reflex have an obvious *raison d'être,* but if you have ever used your withdrawal reflex, you will know how truly unconscious it is—your arm moves before you have even realized that something painful is happening to it.

Reflexes are hardwired into the cord. Many incoming sensory nerves at the back of the cord plug into pathways leading to the motor nerve cells at the front, as we saw in Figure 5.1. Usually they do not directly connect to the motor cell, but instead the message is relayed by one or more intermediaries—internuncial nerve cells, literally "announcing between" cells. Not only can they stimulate muscular activity, but the internuncials can be inhibitory, too—for example, if you need to withdraw a burnt hand, you will need to relax the muscles that hold your arm straight.

If fact, many reflexes have complementary elements to them. The neatest example of this is the crossed extensor reflex—if you tread on something painful with your right foot, then you do a withdrawal reflex with your right leg, just as you can with your arm. This involves contracting the muscles that bend your right leg while relaxing the muscles that straighten it. However, your left leg is not entirely irrelevant, and the pain in your right foot causes the left to straighten—partly to help you stay standing when you lift your right leg, but also to lift your body as far away as possible from the cause of the pain. So your left leg is wired up the opposite way round, with internuncials crossing from the right to the left side of the cord to stimulate leg straightening muscles and inhibit leg bending muscles. Every muscle contraction needs a complementary relaxation and every leg lifted into the air needs another leg firmly planted on the ground.

Yet reflexes are an even more subtle and continuous phenomenon than that. Really, they are the main way we keep our movements smooth and safe. For example, the kneecap or "stretch" reflex is actually the crude outward evidence of an elaborate system of self-protection. Muscles are potentially dangerous things, especially to themselves—remember this next time you wrench your neck—and it is vital that their force is well controlled. The reason that a tap below the kneecap makes your leg twitch is that you are tapping the tendon at the end of the large front muscle of your thigh, the quadriceps. Little sensors in that tendon detect that the muscle is being stretched slightly and feed this information back up in sensory nerves to the cord. The cord's response is to make the muscle contract a little, and the reason it does this is that relaxed muscles get damaged if they are stretched. Of course, we do not spend our time tapping ourselves with hammers, and the stretch is usually caused by something else—either the weight of the body being transferred onto the leg, or the pull of other muscles. As a result, if our muscles lengthen, they always do it in a safe, controlled way, with lots of internal muscle tone to protect them. This one reflex is largely responsible for the graceful way in which our joints can bend first one way and then the other as the muscles around them continually balance and rebalance their pulls.

Not only does the cord smooth and control our movements, it can even initiate rhythmic activity, such as walking and running. We have already mentioned the startling way in which chickens can still run around after being beheaded, but you may be surprised to hear that similar systems exist in mammals. As I have seen in some of my own veterinary cases, some time after severe damage to the spinal cord, cats and dogs can start making paddling movements with their legs, and these are enhanced if the feet are allowed to strike the ground as they would in a healthy, ambling animal. This "running" can occur even if the cord has been completely cut, with no direct connection between the brain and the legs. Similar rhythmicity can even occur to a lesser extent in people—so we now think that our cord carries within it little circuits of cells to coordinate these repetitive movements.

The importance of the cord in the control of movement is made very clear by the strange story of the dreaded disease tetanus, or "lockjaw." Dark, dank, airless places—soil, for example—are excellent sites for growth of the bacterium *Clostridium tetani*. This microbe does not itself attack or live in people or animals, but it has a nasty habit. If it is accidentally spiked into an animal's tissues, it produces the poison tetanospasmin, which climbs up nerves and enters the spinal cord. Tetanospasmin—a tautology meaning "spasm-spasm"— then sticks to inhibitory internuncial nerve cells and stops them from working. The result of this is that there is nothing to control muscle activity, and various parts of the body go into uncontrolled spasm. This can occur anywhere, but in humans it often affects the jaw muscles, causing lockjaw, or "trismus." For some reason, species vary in their susceptibility to the disease—dogs often recover with no more than sedation, whereas in humans tetanus can be fatal if the toxin paralyzes the breathing muscles. That switching off just one component of the internuncial system can cause so much mayhem shows the importance of the cord's internal mechanisms.

One of the main reasons we need to know about the spinal cord is that it is more exposed to injury than the brain. The vulnerability and importance of the spine has been widely realized for millennia, even in cultures where we have no written references to the spinal cord within. An ancient Indian text at least four thousand years old, the *Srimad Bhagwat Mahapuranam,* describes how Krishna cured spinal damage by immobilizing victims and putting them in traction—the supreme and all-compassionate Lord was clearly a practical sort. Instead of being sequestered in a safe, bony box, the cord is mobile, constantly flexing and bending within the sinuous column of bony vertebrae. Not only can the cord be injured by external impacts, but there are also forces within the vertebral column that can wreak havoc. The little dollops of the old notochord that cushion the spaces between our vertebrae, or disks, can burst, or slip, into the canal where the cord lies and destroy it. So, just as the notochord acts as creator of the nervous system, it can also be its destroyer. More Shiva than Krishna, perhaps.

Because the structure of the cord is similar all the way down its length, spinal injuries cause a series of effects that can be easily explained in terms of the cord's anatomy. The most straightforward and disastrous injury is when the cord is completely cut, or "transected," separating a lower fragment of the cord from an upper fragment still connected to the brain. For some reason, the effects of cord transection are initially different in humans than in other animals. For up to six weeks following complete transection, people undergo a process called spinal shock. Virtually nothing at all seems to happen in the isolated lower cord. During spinal shock the nerve cells in that region react to their isolation from the brain by almost ceasing their activity, and the muscles they supply are paralyzed and floppy. We assume that spinal shock is the result of loss of some form of permissive signal from the brain. Although the cord controls and modulates many of our movements, there seems to be some need for the brain to allow the cord to carry out these roles, although this permission from above may be less necessary in other species. As a veterinarian, I see spinal shock as a rather strange phenomenon. Most domestic animals do not display spinal shock much if at all—they instead progress directly to the next, more complex phase.

The symptoms of the subsequent stage of spinal injury depend on the level of the cord affected. They also reflect the different effects of damage to white or grey matter—bundles of axons or clusters of nerve cell bodies. When the cord is transected, the effect on our ability to sense things is fairly straightforward. In a healthy spinal cord, each snippet of sensory information from the skin, muscles, or our insides enters the cord via a nerve cell entering into the back wing of the "spinal butterfly" of grey matter, as illustrated in Figure 5.1. From there, a nerve cell sends its own axon up to the sensory regions of the brain. Cutting the cord may destroy the inflowing axons, the second nerve cell, or the axon ascending up to the brain, but the effect is the same regardless—you cannot feel anything supplied by nerves below the level of the injury. Obviously human patients can tell us what they can feel, but we can also find out what injured animals feel by pinching toes and seeing whether they try to bite us. Spookily, the reflex circuits below the injury may still be intact, and so such patients often still show withdrawal, crossed-extensor, and stretch reflexes.

The effect of cord transection on movement can be complex. First of all, there can be no conscious control of movements by muscles whose nerves arise below the injury. Yet the manner in which these muscles behave varies in surprising ways. Motor commands from nerve cells in the brain travel down the cord in bundles of axons in the white matter which then plug into a nerve cell in the front wing of the grey matter of the spinal butterfly. This cell then sends out its own axon in the front nerve root to reach the appropriate muscle (see Figure 5.1). If the cord is damaged around the level where the nerves flow out to reach a limb, then the nerve cells in the front wing are destroyed. As a result, the target muscle is simply unplugged—causing a distinctively floppy or flaccid paralysis, in which the muscles of the limb have little tone and their reflexes are weak. Also, these muscles shrivel strikingly quickly—within days, in fact.

Conversely, if the cord is cut above the point where the nerves flow out to the limb, then we see very different symptoms. In this case it is not the motor nerve cells in the spinal grey matter that are damaged. Instead the axon bundles from the brain, found in the white matter, are destroyed. These axon bundles would ordinarily course down to reach the motor nerve cells in the grey matter. The spinal nerve cells to the muscles are intact, but they receive no signals from above. Perhaps surprisingly, it appears that the brain controls spinal motor nerve cells with inhibitory rather than stimulatory signals, and the loss of this higher control switches the spinal motor nerve cells into overdrive. The limb goes into a rigid or spastic paralysis and exhibits exaggerated reflexes—a strange sort of paralysis, you might think. So, progressively higher injuries to the cord cause the following symptoms:

Floppy legs, normal arms: injury where nerves to legs exit
  the cord
Rigid legs, normal arms: injury between where nerves to legs
  and nerves to arms exit the cord
Rigid legs, floppy arms: injury where nerves to arms exit the cord
Rigid legs and arms: injury above where nerves to arms exit
  the cord

The differences among all these configurations of injury are not of merely academic interest, as they represent the quickest way for a neurologist to work out which bit of the cord is damaged. You must remember that patients are not wheeled into the emergency room with "spinal cord transection at thoracic segment 12" written on their chest. Instead, clinicians are presented with paralyzed people and must work backward from their symptoms to find out where the injury is. To the experienced neurologist, all these strange, apparently contradictory signs are speaking a language they can interpret into a story of underlying injury. And after all, it is injuries they must treat, not symptoms.

Sometimes only one side of the cord is damaged, so only the limbs on one side are paralyzed. Also, not all damage is sudden—some, like that caused by growing tumors, develops insidiously. Sometimes cord damage may improve with time, whereas in some patients cord damage may be intermittent; all useful information to an expert clinical neurologist. I am constantly impressed by the way doctors have learned how to work out what is going on deep inside our hidden nervous system simply by observing what patients can do. And this ability to mentally peer inside the nervous system will become more impressive as we work our way up to the brain.

So cord damage causes different symptoms, depending on when and how it strikes the nerve cells. But surprisingly, perhaps the commonest cause of cord damage in younger people does not primarily affect nerve cells at all. Multiple sclerosis attacks not only the cord but also the brain in perhaps 3 million people around the world, but it remains an essentially mysterious disease. It can affect anyone, although certain individuals are far more at risk than others. We do not know what causes it, but we have a long list of possibilities. The progress of the disease is extremely unpredictable, to the extent that unpredictability is almost its defining feature. We cannot cure it, but we can often control it, even though we do not know why most of our treatments actually work. All this, without attacking nerve cells.

Nerve cells are not the only cells in the central nervous system. Perversely, they are not even the most numerous. Most of the cells in the brain and spinal cord are spindly, betentacled glial cells. The name "glia" means "glue," reflecting the nineteenth-century theory that these cells served only to bind the nervous system together—a filling to pad the empty spaces. We now know that glia have several important roles, mostly directed toward keeping the environment inside the central nervous system constant and conducive for nerve cells. Some glial cells play yet another important function—they wrap around nerve-cell axons to form a fatty, insulating coating around these "wires" of the nervous system—it is actually this fat that makes white matter white. Not all axons have these glial sheaths around them, but those that do can conduct signals much faster than those that do not.

In multiple sclerosis this fatty coating is attacked by the body's own immune cells. The job of the immune system is to protect the body from foreign invaders, but for reasons that are not always clear, your own immune system can accidentally start to mistake your own tissues for foreign microbes. There are all sorts of different autoimmune diseases, but multiple sclerosis is one of the most common and well known. We are not entirely sure what it is about the glial sheaths that makes them so appetizing to the body's immune cells, but it certainly seems to be glial cells rather than the nerve cells that are the target—initially, at least.

Because multiple sclerosis can affect almost any nerve in the central nervous system, it can cause a very wide variety of symptoms. Damage to the cord can affect not only the signals to and from the limbs, causing problems with standing and walking, but also those controlling the bladder or genitals, leading to incontinence or sexual dysfunction. Sometimes sensation seems to be more affected than movement, and patients may complain of numbness or pain. Damage higher up in the brain can cause more complex effects, including tiredness, depression, dizziness, and, quite commonly, vision problems.

Although these effects make sense now that you know how the nervous system works, it is the unpredictable course of the disease which is harder to explain. Most commonly, sufferers are well most of the time, but they suffer from intermittent flare-ups, presumably as a new region of nervous system is attacked and then scars. We do not know what precipitates these flare-ups, but we would dearly like to know, as each contributes to a gradual deterioration in the patient's ability to function and live independently. Treatment, therefore, is often aimed at preventing these flare-ups. In some patients, the disease can take a very different course. Sometimes the phase of intermittent attacks is followed by a period when damage appears to be slow and steady. Alternatively, some people start off with this slow, creeping form of deterioration and never experience intermittent attacks. And to make life even more confusing, the two phases can mingle together in some sufferers, who undergo a more or less steady, slow decline interspersed with more severe flare-ups from which they only partially recover. It is difficult to explain why the disease can appear in so many different forms, and this variation is even hard to reconcile with the idea that multiple sclerosis is really a single disease at all.

To make matters worse, there are some very strange forces at play in deciding who is afflicted with the disease. Multiple sclerosis is not random—some of us are much more at risk than others. First of all, the condition may be up to three times more common in women. This may seem a striking statistic, but in truth it is one of the less surprising features of the disease. As it turns out, most autoimmune diseases are more common in women, and often much more common. We do not know why women are more at risk, although I speculated about this in my last book, but at least in this context multiple sclerosis is behaving like an "ordinary" autoimmune disease. Another clear trend is that multiple sclerosis is unlikely to start in the very young or very old—85 percent of new cases are diagnosed in people between twenty and fifty years of age. But although uncommon in children, multiple sclerosis occurs often enough to make it difficult to know what these age statistics might be telling us.

There are also some intriguing geographic and ethnic factors that appear to influence who gets the disease and how severe its course will be. Multiple sclerosis is more common in people of North European Caucasian origin than in other ethnic groups, although it does occur in other large subdivisions of the human family tree. Also, there are smaller ethnic communities in which the disease is unheard of. These variations, along with the fact that being related to a sufferer considerably increases your chance of getting multiple sclerosis, have led many to believe that the disease has a genetic basis, even though it is clearly not inherited in a simple way. Superimposed on this issue of inheritance is the challenging finding that the disease is more common in people who live at higher latitudes. Obviously European Caucasians live, on average, farther from the equator than West Africans, but even when this is taken into account, there is something strangely protective about proximity to the equator. For example, multiple sclerosis is more common in the northern states of the United States than in the southern states. There are other diseases that behave like this, but most of them are related to cold environments or dark winters. Why a temperate climate should make you attack your own spinal cord is not so obvious. Yet the effect is clear, and it has even been shown that if you are a northerner, you can reduce your risk by relocating southward before puberty.

As you can imagine from all these unexplained trends and our inability to prevent the disease, we lack a good understanding of what causes it. There seems to be both a genetic component and an environmental one, too. Epidemiologists have reported clusters of cases, but they are difficult to prove statistically, and none has yet been ascribed to a definite cause. Some genes have been implicated in the disease, but they do not appear to work in the straightforward way that we have found genes commonly involved in, for example, familial forms of breast cancer. Also, multiple sclerosis seems to be a peculiarly human problem—veterinarians are not presented with pets and farm animals afflicted with multiple sclerosis—so it is difficult to investigate the causes of the disease in animals. At present we have an uneasy truce in which we assume that the disease results from being born with a genetic susceptibility and later being exposed to an environmental factor—but for now we know neither the genes nor the factor.

Could that environmental event be an infection of some sort? Infections are good candidates because some people are known to be genetically susceptible to certain infections, and some infections occur more in certain environments than others. We even know of autoimmune diseases triggered when the immune system confuses an infectious microbe and some innocent part of the body. In the case of multiple sclerosis, there are several candidates. Viruses can be autoimmune triggers, and most of those fingered are pretty common—measles, herpes, or even canine distemper—but bacteria can set off these diseases as well. One suspect is *Acinetobacter,* a normally harmless bacterium that lives an otherwise unobtrusive life in the soil all around us. It has been claimed that multiple sclerosis sufferers are much more likely to produce antibodies to this bacterium and also that elements of it may bear more than a passing resemblance to molecules on the glial nerve sheath—although findings like this can be notoriously difficult to interpret. But all in all, what with tetanus as well, soil really seems to harbor some antipathy to the spinal cord.

Infections need not always come from outside the body, however. We always think of microbes as being things we catch from our surroundings or from other people, but sometimes we carry them inside us all the time. For example, retroviruses are a group of viruses that infect us in a frighteningly thorough way—incorporating themselves into the very genetic instructions we use to operate our bodies. And it appears that on a few fateful occasions in our evolutionary history, retroviruses have, perhaps accidentally, inserted themselves into the genes in someone's eggs or sperm so that they got passed down to their children. These viral genomes were thenceforth passed down the generations so that all humans now carry several of them in every cell in their body. Over millions of years some of these endogenous retroviruses have become so badly damaged that they are almost certainly functionless, but virologists are increasingly suspicious that some of them do occasionally spring forth from our genes as fragmentary virus particles.

Remarkably, we now have evidence that endogenous retroviruses

are present in the blood and even the nervous system of multiple sclerosis sufferers. Although we cannot rule out the possibility that there
is simply something about the disease that provokes these dormant
viruses into action, there is also a chance that they may actually initiate the autoimmune disease. Although we inherit endogenous retroviruses with all our other genes, they are still foreign invaders at
heart—genetic parasites if you will—and it may be this combination
of the familiar and the foreign that confuses our immune cells. If just
one molecule on our viral passengers were a little like a molecule on
our glial nerve sheaths, all hell could break loose. We do not know
how or when this would happen, nor whether different human populations might have retroviruses more or less similar to their glial
nerve sheaths. But they are at least suitably unpredictable candidates
to be the cause of multiple sclerosis.

Even the management of multiple sclerosis has proved to be controversial—it is difficult to call it "treatment" when we do not really
know what we are treating. When it was realized that the disease was
the result of an overzealous immune system, the first drugs used were
corticosteroids, which have a crude, generalized suppressive effect on
the immune system. Corticosteroids are worthwhile (and very cheap)
drugs if used carefully, but their sheer power is matched by their ability to cause short- and long-term side-effects. The next generation of
drugs used to manage multiple sclerosis were aimed more specifically
at modifying the types of immune response thought to be responsible
for the disease—for example, beta-interferon is a chemical that our
own cells make to modify disease processes going on around them.
We can make artificial interferons, but it is not cheap. In the United
Kingdom beta-interferon has often been at the center of political arguments about what medicines the National Health Service can afford
to dispense.

There is a drug cocktail that has been known for years to alleviate multiple sclerosis. It is easy to imagine how the palliative effects
of smoke from the smoldering leaves of Cannabis sativa were accidentally discovered, but no doubt this furtive illegal discovery of a potential panacea has colored all subsequent political discussion of its ther-
apeutic potential. Scientists are often much more neutral than
politicians on these matters, and a great deal of research is now being

conducted into how cannabis affects the body. Like many naturally occurring drugs, cannabis is a mixture of different but related compounds, in this case called cannabinoids. It just so happens that this species of hemp makes chemicals which bind to our nerve cells and exert dramatic effects—our cells have their own cannabinoid receptors, molecules evolved over millions of years to respond to cannabinoid-like substances that we ourselves produce. We now think that this internal cannabis system is present in most vertebrates, suggesting that it evolved an extremely long time ago—well before Woodstock, at any rate.

One problem with using cannabis as a medical drug is that it has a wide variety of effects, and it is not always easy to work out which of the components of the "wild" drug exert which effects and through which mechanisms in the brain they act. For example, different cannabinoids can exert opposing effects. Also, there are at least two different types of cannabinoid receptor in our body on which they can act. Finally, and perhaps most important, cannabinoids act not only on the nervous system, but also on the immune system as well. This may be why people who use cannabis often report beneficial effects that scientific studies cannot replicate or explain. Cannabis might relieve the perceived symptoms of multiple sclerosis by calming, relaxing, and reducing pain and muscle spasms. It probably also slows the progression of the disease itself by reducing the immune attack on the glial sheaths. So is the calming effect of cannabis just a fortuitous side-effect?

One advantage of using an illegal drug for novel medical purposes is that we already know a great deal about its adverse effects. Obviously cannabis makes people feel rather dopey, and the relaxation could be even more profound if the drug also relieves the exhaustion of chronic pain at the same time. Most patients would probably not wish to feel stoned all the time, but most of the other effects of cannabinoids are widely thought to be relatively harmless. An exception to this is the surprisingly recent realization that cannabis may be a major factor in the development of clinical depression. Clearly we have to balance the possible benefits with the possible risks when we use any drug, but it seems that there are already many people who became willing to risk cannabis side-effects as soon as they started to feel the disability and discomfort of multiple sclerosis go up in smoke.

In some ways, we have come a long way in our perambulations. In terms of distance, we are already well over halfway from the lowest part of the central nervous system, where the horse's tail radiates from the spinal cord, to the highest part of the brain. But simple distance is misleading, and we are nowhere near halfway through the populous cells of the brain and spinal cord, nor halfway through their complexity. Think on this the next time you decapitate a snake. The spinal cord is an excellent preface to the brain, and we are now to travel into more intriguing and mysterious terrain.

6

INTERLUDE

The Worm that Turned (Over)

In central Paris, there is a green island of calm hemmed in by the bustle of the *cinquième arrondissement*. The Jardin des Plantes is a place to sit, relax, and watch the world go by, but it is also somewhere you can go back in time. To step into the zoological gallery of its Muséum national d'Histoire naturelle is to step into the way museums used to be. I am just about old enough to remember the time before the advent of the push-buttons and interactive video displays that now infest many of Britain's major science museums—a time when specimens were allowed to speak for themselves, rather than being stored away in basements and replaced with audio commentaries and animatronic recreations. I well remember as a child wandering through the fossil fish galleries of London's Natural History Museum and, as I looked at the crumpled, petrified inmates, being struck for the first time by a sense of just how very old the world must be. The fish are now filed away out of sight and replaced by something altogether less challenging, but I still remember what they told me.

The French clearly have more respect for the way that impressive things can impress you, even though you might not know exactly what they are. The zoological gallery in Paris is simply a hall full of hundreds of animal skeletons and various biological specimens pickled in glass jars. There are written explanations of the displays, but they are mercifully brief, charmingly faded, and often reassuringly out of date. They are really telling you not to read them at all but to look at the specimens themselves. Does the elephant not look just as majestic in death? Do the giant lizards not still fill us with a sense of unease? Is it not amazing how you can construct the skeletons of a giraffe and a guinea pig with essentially the same bones? Is there not a sense of the unity of nature in this place? N'est-ce pas?

Much of this wonderful, morbid menagerie was assembled under the guidance of one man, Étienne Geoffroy Saint-Hilaire, who was the professor at the Jardin from 1808 to his death in 1844. Geoffroy was the best kind of scientist—an argumentative, error-prone visionary who could so easily have ploughed an equally controversial furrow in many other fields of human endeavor. Born in 1772 in Étamps, his early life hints at a certain indecisiveness regarding vocation and belief—he was made a canon of the church at the age of 16, but by the time he received a law degree at 19, he was a committed deist, no longer believing in the revelations of orthodox Christianity. At that time, the deist re-evaluation of the nature of God was central in the thinking of revolutionary France and may be one reason why that state is now so adamantly secular. Certainly the young Geoffroy was active in the Revolution, building up revolutionary cachet that was later to allow him to negotiate the freedom of friends and colleagues during the Reign of Terror, the convulsion of inward-looking recrimination and paranoia that racked the nation in 1793 and 1794.

While the political maelstrom of the Revolution swirled around him, Geoffroy was developing a love for the study of animal structure, and an astonishment at the ways in which apparently different animal types could share so many structural features. We have already alluded in Chapter 4 to the remarkable similarities among the brains of crawling, swimming, running, flying, and speaking creatures, and all these wonders and the questions they posed fed Geoffroy's imagination. He gradually became obsessed with the search for the "archetype"—the conceptual basic animal plan on which all these variations are superimposed.

To most modern biologists, Geoffroy is of interest because he epitomizes the enlightened view of animal types before Darwin's master stroke. He toyed with many of the ideas that would connect together to form the theory of natural selection, without actually taking that final step. He speculated that animal species might change over time, and that this may be the reason why disparate animals share so many features for no obvious functional reason. Why should whales have arm bones in their flippers, for example? He even suggested that animals' characteristics might be inherited by their offspring and that the environment might in some way influence which ones are inherited. No surprise that he was later to be quoted in *On the Origin of Species.*

A turning point in Geoffroy's life came when he was appointed as one of Napoleon's scientific staff on his military-cultural-scientific campaign in Egypt. Though Geoffroy was an active member of the expedition, his tour of duty in Africa drew him away from the intellectual ferment of Paris for three years in his mid-twenties, and he seemed strangely isolated on his return. Although his greatest works were still to come, he was a different man after his isolation in Egypt—detached and unwilling to countenance the ideas of those with whom he disagreed. In some ways this was liberating as it meant he could freely pursue his beloved archetypes, but it also meant he was wont to stretch his ideas too far. He slowly edged toward the idea that there was a common body plan that united not just the vertebrates, but all animals. He is often quoted as saying, "There is only one animal." One blueprint for worms, gnats, starfish, and people—a kind of animal *fraternité,* perhaps.

There were considerable problems with developing the idea that all animals are based on the same plan, not least of which were the apparently irreconcilable differences between the bodily arrangement of the two most obvious groups, the insects and the vertebrates. Insects, and their tough-cuticled, jointy-legged cousins the spiders, crustaceans, and centipedes, have a gut tube that runs from their mouth to their anus along their back (Figure 6.1). Atop this is a network of vessels which pump a fluid that can loosely be termed "blood" toward the front of the animal. Most important for us, insects have a nerve cord that runs from head to tail along their belly. Not very much like vertebrates, you may think. It took someone like Geoffroy to do something of almost childlike simplicity: He turned the insect diagram upside down. Now it looked more like a vertebrate—gut below, heart pumping blood below that, and nerve cord running along the

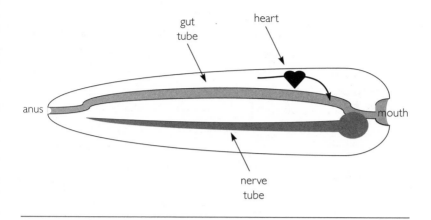

Figure 6.1. A proposed body plan for the ancestor of many multicellular animals. Insects, for example, have a nerve tube (central nervous system) running along their belly and a gut tube running along their top edge. Above the front part of the gut tube is the heart, which collects blood from behind and then pumps it forward and down. If you turn the page upside down, is what you see the body plan of a backboned animal?

top. He did it in 1822, and you can try it now with the illustration in this book.

Was it really true that vertebrates are simply the insect body plan upside down? And how in the world could that happen? Geoffroy had a ridiculously simple explanation. He argued that at some point in the distant past there was a little wormy creature with an internal heart/gut/nerve tube arrangement similar to that of, let us say for sake of argument, an insect. Some of the descendants of this wormy thing sprouted legs and eventually became insects. But others rolled over sideways and started swimming on their backs—and then in their new, inverted posture developed fins and legs as they started the slow trek toward becoming today's vertebrates. Thus, by dint of a kind of permanent backstroke, our family of animals arose.

As you can imagine, this idea that vertebrates are inverted insects received acclaim and derision in equal measure. It was so simple that everyone could understand it, and everyone had an opinion. Unfortunately, Geoffroy did not do himself any favors with some of his more unlikely elaborations to the story—for example, claiming that the

hard exoskeleton of insects is the equivalent of the vertebrae that protect our spinal cord, so that insects are thus condemned to a life of
walking around on the tips of their ribs. Anyway, the popularity of the
inverted-insect idea has waxed and waned ever since, although in the
twentieth century the "wane" seemed to be getting pretty permanent.
The theory seemed destined to be consigned to the dustbin of the history of science. But that was until the advent of molecular biology in
the late twentieth century.

Four or so decades of hard slog have now taken us to a position
where we can actually watch genetic instructions winking on and off
in developing embryos. As a result, we can watch how the interactions between all these genes convert an apparently amorphous fertilized egg into a complex, ordered organism. We started by deciphering
the developmental genetics of fruit flies and roundworms, but the
vanguard has moved on to animals more closely related to humans:
zebrafish, frogs, and mice. I was studying for my bachelor's degree in
the year that scientists first discovered that the genes that coordinate
the development of a mouse are almost identical to those that pattern
a fly. Suddenly it seemed that an overarching mechanism for constructing all animal types was lying hidden behind our differences.

Soon this long-hidden world of embryonic genes was to provide
unexpected evidence that, after all, Saint-Hilaire's inverted insect idea
might not be so ridiculous after all. For example, the patterns drawn
by the genes that control the formation of the nervous system are
both so complex and so similar in insects and vertebrates that it is difficult to see how they could not be derived from the same ancient
structure, albeit in a "right way up" and an inverted form. And the
genes that control the development of the heart of the two groups of
animals told the same story, even though their hearts are oriented in
opposite directions. The same is true for the genes that code for
genes, and the segmentation genes, and the genes encoding legs.

The most dramatic support of Geoffroy's idea came in the last two
decades of the twentieth century with the discovery of the genes that
control which side of the body becomes the back and which side becomes the belly. Only once these genes have done their job does an
embryo know which way up it is, and in a delightful convergence of
events the genes that define a mouse's belly also define a fly's back.

And vice versa, of course. Surely the realization that the inversion of fly into mouse is a reflection of the very mechanisms that establish the orientation of these animals suggests that Geoffroy was right after all? Are we, after all, the worm that turned over?

Despite this impressive contemporary evidence in support of our hero *savant*, there are still those who doubt the inversion idea. They argue that there are other ways in which an ancestral creature could have organized itself into both an insect and a vertebrate without any need to flip over, and these theories usually relate to the origins of the nervous system. For example, there is no rule that says that the nervous system must always have been concentrated in a tube or cord along the top or bottom of the body. An alternative hypothesis is that the ancestral nervous system was originally a diffuse network of nerve cells pervading the whole body and that this mesh slowly condensed into the central nervous system. By some quirk of evolutionary history, the nerves condensed above the gut in some animals and these gave rise to vertebrates, while in others the nervous system condensed below the gut, giving rise to insects and their kin. And maybe the genes that now control the orientation of the body acquired their present roles during this ancient process of nervous system condensation—with some genes drawing it toward the back in some creatures but drawing it toward the belly in others.

A competing theory is that the ancestral nervous system was not in fact diffuse, but instead was already concentrated into two symmetrical lines of cells running along the left and right sides of our forebears. According to this scenario, the two nerve cords are connected up to lines of cells running along the organism's flanks. These two nerve cords may have subsequently crept either "down" to fuse into the insect belly-nerve-cord or "up" to fuse into the vertebrate back-nerve-cord. Either way, the gut is left running along the middle of the animal, with the nervous system above or below it.

To me, Geoffroy's idea remains a compelling one, because it has emboldened us to think about the distant origins of the vertebrate nervous system and its original place in the body. We are thinking back to the unnervingly ancient Precambrian era—maybe six to eight hundred million years ago—certainly long before those squashed fish I used to peruse in the museum. It is a frustrating but tantalizing business—we have almost no fossil record of what was going on at this time. But we suspect that the construction plans for our bodies were being written right then, in an era now hidden from us by the march of time.

Another feature that many animals evolved in these ancient seas was the very seat of the brain—the head itself. Heads are such useful things that they have evolved again and again in different groups— vertebrates, insects, molluscs, and several other groups have a "head end." If you are going to be a mobile organism, moving purposefully in a particular direction, it makes sense to cluster certain organs in the leading end of your body—your mouth, for example, or some sense organs. And with his "servants and guards of the great king" analogy, Galen has already told us what happens when you have lots of sense organs in one place—a brain forms to analyze all the incoming information. However, there is disagreement about this—some biologists think that the brain actually went to the front end of vertebrates to control what was allowed to enter the mouth, rather than to plug into sense organs. Maybe they are right—and certainly the fact that it is the front of the gut rather than the notochord which induces the development of the front of our brain lends credence to this idea. The front end of the brain is special, and it is the mouth that makes it so.

The formation of the head, much discussed in Geoffroy's time, has become a hot topic in the modern science of evolutionary developmental biology. Among the genes we have discovered which control embryonic development are those that specify the head and tail ends of the body, and once more we find similar genes controlling the head-to-tail arrangement of both insects and vertebrates. And in both groups, the clustering of organs into the head continues to this day. It even has a name: "cephalization." As certain vertebrate groups have evolved, they have crammed more and more functions into their heads and brains. We humans are perhaps the most cephalized animals of all, and almost all the original functions of our spinal cord have been subsumed into our brain. This is probably why, for example, we show such weak movements when our cord is disconnected from the brain compared to snakes and chickens. It may also be why spinal shock is so severe and long-lasting in humans—the isolated human cord is a relatively bereft and inadequate thing compared to that of cats and dogs. In us, all the action is higher up. We have cephalized to the extreme. We are, in the scheme of things, the most headstrong creatures of all.

I think Geoffroy would have been happy. He received a lot of criticism for his theories during his lifetime, but maybe some of them have been proved right. He also made it possible for us to think back to the time when our bodies were forming—a time of which we have no physical record. Best of all, we even have the single archetype he sought. It is not an animal, but a common toolbox of interacting genes that any animal can use to make itself—including any of those beasts in the glass cabinets of the Muséum national d'Histoire naturelle.

II

AN ASSAULT ON
THE SENSES

The brain stem itinerary:
Plunging from the zonules of Zinn
into the inner labyrinth

It was a case of finding out how the roots and branches of these trees terminate,
in that forest so dense that there are no spaces in it,
so that the trunks, branches, and leaves touch everywhere.

—Santiago Ramón y Cajal, *Recuerdos de mi Vida*

7

A FOREST SO DENSE

The New Anatomy of Santiago Ramón y Cajal

It had worked! A loud explosion echoed around the valley and a satisfyingly enormous cloud of grey smoke enveloped the excited boys. The blast had almost knocked them off the orchard wall.

The thick wooden barrel of the cannon had remained intact, but the tacks and cobblestones that the boys had selected as projectiles had vanished. It was soon to become apparent where they had gone. The billowing plume slowly cleared and Santiago and his comrades were greeted by the wonderful sight of the devastation they had exacted on their target. The neighbor's beautiful new garden gate was torn into tiny splinters, and the proud gateway was now filled with nothing but its angry owner, soon to launch himself toward his tormentors with immense ferocity. The cobblestones were soon flying in the other direction, albeit with slightly less speed, but a few of them clipped the boys as they fled into the Iberian afternoon.

One boy, Santiago, was caught, and as the manufacturer of the homemade artillery, he had to bear the brunt of the punishment. The mayor was soon informed of his explosive misdemeanor and the eleven-year-old was soon locked in the local prison cell, a stinking and cockroach-infested pit into which the locals peered and yelled through a small barred window. His father was fully in favor of this punishment, and the boy was allowed to languish in the small rural prison for four days and nights with neither food nor water. Even the cockroaches had started to look succulent by the time his mother and aunts started to smuggle meat and pies to the captive. Some youngsters learn slowly, however, and on his eventual release Santiago would redouble his attempts at cannonry, although the next weapon was to destroy itself rather than any intended target. It also sent a sliver of brass deep into his eye, tearing through his iris to leave a permanent scar.

Such were the early days of the great Santiago Ramón y Cajal, the finest scientist Spain has ever produced. The son of a provincial surgeon, he was born on May 1, 1852, in Petilla, a little backwater in Aragon in the north of the country. In later life he seemed almost apologetic about his rural origins, but I cannot help feeling that Ramón y Cajal's untrammeled childhood was what made him a scientist. Young Santiago no doubt worked a great deal of frustration, mischievousness, and violence out of his system and is now remembered as one of the most meticulous, benign, and gracious scientists who has ever lived. It is probably for these reasons, even more than the immense insights he left us, that he is so fondly commemorated in Spain. Even before I knew who he was, I had noticed that his country boasts many *Calle Ramón y Cajal*—Ramón y Cajal streets. Britain's towns do not boast streets named for neurohistologists and neither, I suspect, do cities in North America.

Santiago's father's occupation meant frequent moves—to Larrés at two years old, Luna at three, Ayerbe at eight, and hated Jaca at ten. To read his colorful autobiography, it sounds as if he left a trail of destruction behind him. Despite his father's calm counsel, the boy's erratic and rebellious nature was almost his undoing. Really, it is fortunate that he survived to his illustrious adulthood. At the age of four, for example, he took it upon himself to beat a horse, not entirely understanding the disparity in strength between him and the animal. His victim did not wait long to kick the child soundly in the middle of his forehead, knocking him unconscious for some days, and creating a severe wound that bled profusely.

Even a youthful interest in ornithology became a life-threatening pursuit in the hands of Santiago. Not content with having his hand

severely lacerated by a family of rats that had resented his groping intrusion into the magpie's nest in which they had set up home, Santiago decided that he wished to view an eagle's nest halfway down an
especially smooth cliff in the Sierra de Linás. It was a simple enough
matter to leap downward from ledge to ledge until he neared the eyrie
to see the squawking chicks, but Santiago then realized that there
were no ledges farther down, and the ledges above were separated by
wide, featureless expanses of rock that offered no handholds. Near to
collapse with hunger, thirst, and sunstroke, he desperately slashed at
the rock face with his penknife and found it surprisingly soft. Using
this tool, Santiago was able to scratch out tiny cracks into which he
could wedge his fingertips, and by this slow and painful means was
able to edge his way up to the top of the cliff.

What was Santiago's father to do? Justo Ramón Casasús was a well-
meaning man, although his errant son's behavior often left him bewildered and at a loss. It was clear that the lad was an accomplished artist, as he had been producing wonderful drawings from an early age.
Yet Ramón Senior was of a practical bent and he was determined that
his son should receive a good education, or failing that, study for
trade. As a teenager, he was sent to a strict school in Jaca that was
hardly suitable for a boy of his temperament. The curriculum was
based on classical languages and learning by rote, and this instilled in
Santiago a hatred of both. He was often at loggerheads with his teach-
ers and this usually escalated into punishment by beating and starvation. His father became more and more desperate to find something
to occupy the lad—apprenticing him first as a cobbler and then as a
barber before finally stumbling upon a macabre solution.

Grave robbing may not seem the obvious way to bring a youth onto
the straight and narrow, but in Santiago's case it did the trick. The
Catholic faith has a very open and practical attitude to death, and a
country surgeon in nineteenth-century Spain must have been about
as pragmatic in this respect as one can imagine. Years earlier, Justo
Ramón Casasús had taught his son French as the two sat in a cave, so
why should he not now be able to inspire the boy by teaching him
anatomy in the garden shed? One night, father and son stole away to
the nearest graveyard and found, in a hollow, piles of bones exposed

by rainstorms or heaved from the ground to make space for new corpses. Cajal was later to enthuse,

> Deeply was I impressed by finding and examining these human relics! In the pale glow of the lantern of the night, those skulls half covered with fine stones, with irreverent thistles and nettles clambering over them, seemed something like the wreck of a ship cast upon the shore.

Father and son picked through the bones and hurried away with a collection of the finest, and to Santiago's admitted relief no divine punishment was to strike them.

This heart-warming scene of nocturnal exhumation was the spark that lit Santiago's intellectual fire. His father began to teach him the knobbly terrain of trochanters and tuberosities that covers the bones—the same terrain that, in canine form, I spend most of my time teaching to students—and the boy became gripped by the ancient, gnarled shapes that were now in front of him. He started to draw the bones and soon fell into the well of anatomical fascination that seems to swallow some of us. The structure of bodies was to be his lifelong fascination, although he did not yet realize that the main object of that fascination was to be the grey stuff long rotted out of the skulls they had stuffed into their sacks on their nocturnal forays into the graveyard.

He was soon excelling at other sciences and mathematics too, with a particular love of geometry and algebra. Instead of a life repairing shoes, he went to medical school in Zaragosa, where he tutored others in anatomy when he was still a student. After enlisting in the army at twenty-two as a medical officer, he traveled far and wide, including a long stint in Cuba where he caught malaria. Rarely can catching malaria be described as lucky, but for Cajal it led to an honorable discharge and an academic post at Zaragosa. He was soon director of the anatomical museum there and married Doña Silvería Fañanás García, who was to bear him seven children. Happy at home and work, Ramón y Cajal was the antithesis of the tortured genius, although he was careful to control what we would now call the work-life balance, later stating that "the children of the flesh did not smother the children of the mind." The children of the mind clearly flourished, and by the age of thirty-one he was professor of anatomy at Valencia. Ramón y Cajal's mind had been set free by the bones.

Perhaps surprisingly, Cajal was not to become famous for working on parts of the body visible to his artistic naked eye. Instead he was to visit the microscopic world of the brain. By this time, the gross structure of the brain had been thoroughly mapped, but its internal wiring remained essentially mysterious. This was not because of any inadequacy of contemporary microscopes; it was simply that the brain seemed frustratingly amorphous when viewed through them. For two centuries, microscopists had used stains to emphasize certain elements of their biological specimens. The inventor of the microscope, Anton van Leeuwenhoek himself, had stained nerves with saffron and brandy, but since that time the tangled mass of nerve fibers in the brain had remained disappointingly impenetrable.

Cajal was painfully aware of this when he first turned his microscope on thin slices of the brain. With the staining techniques available to him, ghostly apparitions of cells speckled the brain, but he could not see how they interconnected. Neuroanatomists elsewhere in Europe claimed that brain cells were all in physical contact with each other, interconnected and fused into an enormous net of merged cells, with thoughts sloshing backward and forward through their collective mass. Yet it seemed unlikely to Cajal that such a homogenized mesh was the basis of our intellect—it ran counter to all that anatomists had shown about the heterogeneity of the structure of the brain. However, he had no way to prove otherwise.

His breakthrough came in 1887 when a colleague first introduced him to a new staining technique invented by Camillo Golgi, a professor at the University of Pavia in Lombardy. Golgi had used silver to highlight nerve cells within brain tissue, although the technique was rather variable in its effectiveness. In the skillful hands of Cajal, this variability was turned to great effect. It had originally seemed frustrating that Golgi's method only picked out a small percentage of nerve cells, leaving most unstained, but Cajal realized that in the tangle of the central nervous system, this high failure rate could be an advantage. It might mean that he could see the neural wood for the trees, as it were. Cajal standardized the technique to produce reliably sparse black staining, and then applied it to thicker slices of brain to allow him better to follow the meandering tendrils of the brain cells. He now had his main instrument of research, and he was to spend the rest of his life mapping the brains of humans, mammals, birds, reptiles, fish, and invertebrates. At last he had eyes to see, and they were the eyes of an artist. Even today, with spectacular imaging technology available to us, Cajal's images of the inner structure of the brain remain a remarkable artistic and technical achievement. We are lucky that Cajal was working before the age of scientific photography, or we would have been denied the beautifully simple drawings that to this day are the best representations of the community of nerve cells in our head.

The first tenet of nineteenth-century neurobiology that he challenged was the idea that the nerve cells in the brain are fused into an interconnected mass of protoplasm through which our mental processes ebb and flow. As Cajal carefully recorded the arrangement of the blackened nerve fibers he saw through his microscope, he realized that he now had clear evidence that the nerve cells do not simply fuse together where they meet. Instead, when two nerve cells interact, there are usually little swellings at the tips of the twigs of one of the cells.

We now call these swellings "synaptic boutons," or more recently the depressingly Anglo-Saxon "synaptic knobs." Electron microscopy has now shown us their internal working parts—but long before that the boutons were enough evidence for Cajal to claim that nerve cells may interact, but they do not meld together. Everywhere he looked, nerve cells were in intimate apposition, but separate. The brain is not a merged mass of billions of cells, it is a community of discrete separate entities. Other scientists went on to use his results to elevate a new scientific theory—the "neuron doctrine"—in which the central indivisible unit of the brain is the nerve cell. And to reflect its new elevated status, the nerve cell was glorified with the suitably elemental-sounding name "neuron."

The neuron doctrine has triumphed. It now lies at the core of our understanding of the function of the brain, even though it is largely based on what Cajal told us about the structure of the organ. The idea of neuron as elemental unit of the brain was soon to be extended to include the neuron as the unit of information processing and the neuron as the unit of embryonic brain development. This is the reason why most neuroscientists think that the functioning of our apparently subtle minds will eventually be explained by studying the interactions among these simple, individual cells—to the modern mind, cells and their interconnections are all there is. And remarkably, Cajal worked most of this out himself. He was unhesitatingly gracious about giving credit to others. Although he did not concoct the word "neuron" himself, nearly everything else in modern neurobiology can be traced back to his ideas.

An important ramification of neuron doctrine was that the brain could have discrete, separate pathways for different functions. With the integrity of his silver-impregnated nerve cells in his mind's eye, Cajal was able to look more clearly than anyone before him at the nervous system. He was especially interested in the special sense organs—the eyes, ears, and nose—because they had an obvious function. For example, it was obvious that somewhere in the retina at the back of the eye lay the cells able to respond to light and to send information back to the brain. Cajal could see that there are two different types of photoreceptor, which we now call rods and cones. Others had already speculated that one cell type responds to dim light and the other to bright light. Cajal showed that whereas several rods or cones might all feed into the same pathway back to the brain, the pathways of rods remained entirely separate from those of cones, even though they ran alongside each other all the way along the optic nerve back to the brain. The discrete, unmerged nature of neurons meant that isolated pathways could be kept apart for long distances. After all, why bother to discriminate between dim and bright information if you are just going to mix it all into a blur?

Cajal also realized that this idea of separate pathways could be extended to the macroscopic level. It had been known for centuries that the optic nerves seem to cross at an X-shaped optic chiasm under the front of the brain ("chi" is the Greek letter "χ"). Cajal now showed that only half of the nerve fibers in the human optic nerve cross to the other side, in a manner which means that the two sides of the brains "see" very different views of the world. This segregation of the information in the two sides of the brain also means that there are actually very few places where information can cross from the right cerebral hemisphere to the left and vice versa. Cajal had rejected the idea that information can slosh about as it likes in the brain, but these links between the two hemispheres now seemed like real bottlenecks between our two otherwise separate brains. We will return to this crucial separation of left and right later.

Cajal was also important in determining how the brain forms in the first place. Unlike most organs, which only have to grow into the correct shape, the brain must also form all the right internal connections. Cajal showed that the basic unit of all this interconnection is the neuron. In a detailed series of experiments he showed how nerve cells grow. He observed processes grow out from the nerve cell body, tipped by an irregular, flared structure called the growth cone, which looks as if it is hunting out the eventual destination of the processes. We now know that this is exactly what the growth cone is doing. The growth cone seems to react to local chemical cues when it reaches that destination, because it then disappears and is replaced by synaptic boutons. Repeated billions of times, this is how the developing brain wires itself—swarms of neuronal processes tracking about the place, past and around each other, until their quest is over.

This swarming, embryonic brain is in itself an impressive enough concept, but Cajal extended it even further. First, he speculated that intelligence is a reflection not of the number and organization of neurons themselves, but of the number and organization of their connections. Each brain cell has many connections to other cells, and so there are many times more connections than cells in the brain—and this today forms the basis of what little we understand about brain size and intellect. Second, he also suggested that although most neuron development was complete by birth or infancy, the growth, regrowth, or strengthening of connections among neurons could be the physical basis of learning and memory. And once again he has turned out to be substantially correct. It is stunning how much of a foundation Cajal laid for modern neuroscience with only some slices of brain, a bottle of silver, and a pen.

While working on the nerve cells in sense organs, Cajal made another crucial leap forward in his ideas about the flow of information in the brain. He proposed that neurons can only conduct information in one direction. Impulses do not meander across neurons in a disordered manner. Instead each one has an input and an output. Because neurons are directional, they cannot be wired up backward—an electrician would think of them as diodes. Even before Cajal's experiments, it had been noticed that neurons often have two very dissimilar ends (see the cell on the right in Figure 7.1). One is a radiating fan of tiny filaments called dendrites (because they look like the branches of a tree, or "dendron") while the other is usually a single, large cord called an axon ("axis"), although the axon may branch a little before it terminates in its synaptic boutons. Cajal noticed that the neurons in the eyes are always oriented the same way, with the dendrites nearer the light and the axons nearer the brain. He also showed that nerve cells in the smell pathway are oriented in precisely the same way (although he judiciously ignored the hearing pathway for reasons we will mention later).

Based on these observations, he proposed his grand Law of Dynamic Polarization, that a nerve cell receives impulses through its dendrites, collates them, and then sends impulses out through its axon, usually to the waiting dendrites of the next nerve cell along the pathway. This theory gave scientists a powerful conceptual tool. Now they could see how neurons with a sense of direction can be constructed into a system that takes in sensory information, analyzes it, and converts it into meaningful responses. We see an apple on the branch, and reach out to pick it. The directionality of nerve cells has proved to be one of their fundamental properties. Inspired by what he could learn from sense organs in vertebrates, Cajal also traveled to Santander and Las Palmas to study the huge eyes of squid, and he found that exactly the same principle holds true. No matter how different animals are, whether vertebrate or invertebrate, their senses have to detect the same things, and the nerve cells that convey their sensations are very similar.

So, just as Galen had established the function of the brain by studying its gross anatomy, Santiago Ramón y Cajal now illumined its inner workings. His theories—broad in their scope and easy to test—are so fundamental to modern neuroscience that they are often taken for granted. That might have been enough for one man, but Cajal was ambitious, intent on changing the image of Spain as an intellectually backward country. Cajal did more than any other to change that

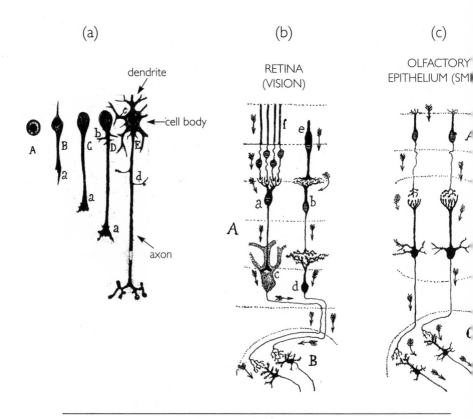

(a)   (b)   (c)

dendrite

RETINA
(VISION)

OLFACTORY
EPITHELIUM (SM

cell body

axon

Figure 7.1. Three drawings of nerve cells by Santiago Ramón y Cajal. (a) Episodes from the life of a growing embryonic neuron. The nerve cell is extending its main branch, or axon, down the page. During this extension process the axon is tipped by the questing growth cone. Only once the axon reaches its intended target does the axon take on its final shape. (b & c) Diagrams of the nerve cells that convey visual information from the light-sensitive cells of the eye and from the odor-sensitive cells of the nose to the brain. Three of Cajal's ideas about the nervous system are evident. First, the nerve cells do not physically touch each other. Second, the flow of information can be contained within discrete pathways, even though many such pathways lie in close physical proximity. Third, as his arrows indicate, nerve cells transmit information in one direction only, from the multi-branched dendritic end to the large, single-trunked axon end.

view. He set up his own journal, the *Revista Trimestral de Histología Normal y Patológica,* to allow Spanish neuroscientists to disseminate their work, but even then became frustrated by the limitations his language placed upon him. He translated his papers into French to widen their readership and started to join German scientific societies to have his ideas heard. It was here that the sheer quality and insight of Cajal's work started to dominate European biology. The tables were turned and German biologists were learning Spanish so they could read his work.

From then on, Cajal's place in history was assured. He worked harder and more productively than ever before, and did so until his death in 1934. Unsurprisingly, he was awarded a Nobel Prize in 1906. For reasons that remain unclear to this day, the man with whom he shared the prize, Camillo Golgi, used the ceremony as an opportunity to attack Cajal's theories and scientific respectability, but Cajal remained characteristically polite throughout. Respected around the world and idolized in his own country, he had revealed the indivisible elements of the brain, explained how they connect, and hypothesized how they channel our thoughts. And much of this he showed us by studying the senses—the frontiers of the brain that bring the outside world in. The senses still represent the easiest route for us to study the brain because, in short, we know from the outset what they are supposed to be doing. We have already seen that the sense organs probably determine the location of the brain in the vertebrate body plan, and so it will come as no surprise to us to learn that the senses have also determined its form.

8

THE LITTLE FISH WHO NEVER GREW UP

The Origins of the Ear

We are now moving on from looking at anatomy—what we already know about the brain—to evolution—the thing that we will never really know for sure about the brain. This story does not take long to tell, but it recounts the events of hundreds of millions of years. The story of ears is the story of whence we came and the lives our ancestors lived. It also tells us how we got our hindbrain.

About 800 million years ago, we were still at the bottom of the vertebrate family tree. Many evolutionary biologists think that our active, swimming vertebrate ancestors evolved from altogether more sedentary life forms consisting of a serrated throat anchored to a stone in some ancient abyss. If modern-day animals of this type (such as sea squirts) are anything to go by, however, our immobile forebears probably had a clever way of spreading their progeny through the environment to find that perfect anchor-stone—they had swimming larvae. Their newly hatched offspring used strips of segmented muscle wiggling a sinuous notochord to move from place to place—to swim, in fact, in a very fishlike manner. They still had a serrated throat at their front end to pick up the occasional snack while on the move, but now they also had a notochordal outboard motor at the back (Figure 8.1).

One day, maybe one of these larvae decided never to grow up. It

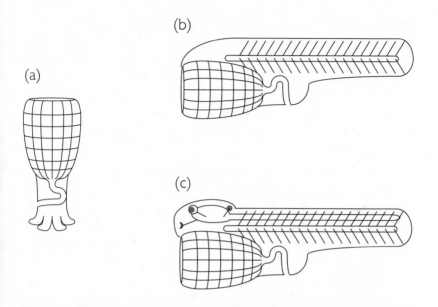

Figure 8.1. Three likely stages in the evolution of the vertebrate body and nervous system. *(a)* A sedentary organism with a serrated, basket-like throat can filter food ·particles from the water. *(b)* A mobile organism that moves its throat through the water with a tail. The tail has a central, gel-filled notochord that flexes from side to side by means of a series of muscular segments. This animal was probably very similar to the larval stage of the previous creature. *(c)* With mobility came the need to co-ordinate movement and react to the environment. A tubular brain and nervous system runs along the top edge of this animal, and organs of special sense appear around the brain. This stage is not unlike a simple fish or tadpole.

had a way to eat. It was mobile. Why glue your bottom to a stone when you can cruise around, hunt, and find a mate? It seems quite likely that this is how the vertebrates, or perhaps more correctly the chordates, got started. They failed to metamorphose and instead became sexually active mobile infants who could procreate to produce more sexual infants. This is a process called pedomorphosis. You may not like the sound of it, but I am afraid that is what we are.

Swimming uses a great deal of energy, however, and our first mobile ancestors had to make some changes. A mobile lifestyle gives you choices. You can seek out food where it is most plentiful and you can take deliberate steps to avoid big, unfriendly animals that want to eat you. But to do these things you need sense organs, and we think that the time when chordates committed to free-living existence was also the time when they started to develop such organs. And when they grew their sense organs, they needed to grow a brain to deal with the flood of incoming sensory information.

One of the things chordates urgently needed to sense was movement. When you are glued to a rock, there is no point in detecting movement because there is nothing you can do about it anyway. But a bold, active, swimming vertebrate has to feel the water coursing over its flanks to tell it where it is going, where the currents are taking it, and whether prey or predators are darting past. We think that the way early vertebrates did this was to have a line of ciliated cells running down each flank—not unlike some of the hypothetical animals mentioned in the preceding Interlude. These cells bore one or more spindly cilia or "hairs" that wafted in the seawater, detecting water flow, compression by nearby objects, and perhaps low-frequency vibrations, but probably not the rapid pressure waves of sound. And the hair cells connected to the back of the brain by a large nerve.

How do we know this? Because this is exactly the system used by fish today, and we see evidence of it in all the fossil fish we have ever disinterred. You have probably noticed that fish, such as goldfish, have a seam passing all the way along each flank. This is the lateral line, and it contains clusters of hairy cells called neuromasts that detect just these things—flow, compression, and vibration. The lateral line is also present in tadpoles for that matter, but it disappears as they turn into frogs. In many modern fish the neuromasts are enclosed in a sealed tunnel to enhance their sensitivity, and there are usually extra lateral lines running invisibly over the face, but the principle remains the same. The neuromasts send their information via thick nerves into the hindbrain, where it is processed in cell clusters that are the direct antecedents of the regions in your own brain that analyze sound and balance. We may have discarded our lateral line now that we do not swim so much, but the wiring left behind was certainly useful.

Our ancient swimming ancestors also needed to know their own orientation. You can only make purposeful swimming movements if you know which way up you are and know the difference between

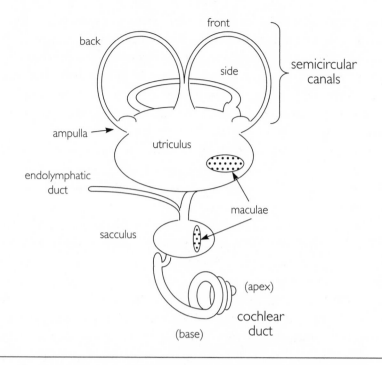

Figure 8.2. The left inner ear labyrinth. The labyrinth is a complex hollow structure. At the top are three semicircular canals that detect rotation. In the middle are the cavities of the utriculus and the sacculus, which detect gravity and acceleration. In vertebrates that can hear, the site of sound detection varies, but in mammals it occurs in the coiled cochlea at the bottom of the labyrinth.

swimming in straight lines and swimming in circles. Working out if you are upright or if you are spinning around cannot, as it happens, be decided by contact with the surrounding water. Early in our piscine evolution the head internalized a cavity lined by neuromasts. This effectively isolated these sensitive cells from the surrounding eddies and fashioned an organ able to detect gravity and rotation. This organ is hollow, complex in shape, and lined on the inside with hair cells. Because of its complexity, it is called the labyrinth. I have drawn the human labyrinth (Figure 8.2), but the labyrinths of other vertebrates are fairly similar, at the top end at least. A more prosaic name for this structure is the inner ear.

Scientists are confident that our labyrinth originally evolved as an internalized bubble of surface hair cells because of two lines of evidence. The first is that there are some living vertebrates with their labyrinths still only partway through the internalization process. Cartilaginous fish—sharks and rays—have a labyrinth similar to ours, but with one difference. In Figure 8.2 you can see that there is a cul-de-sac in the human labyrinth called the endolymphatic duct, which points toward the brain and terminates just outside the outer layer of the meninges. Yet in some cartilaginous fish, the duct points outward toward the surface of the head and actually opens out onto the outside world. So in these fish seawater enters the ducts and fills the labyrinths.

The second piece of evidence for the internalized labyrinth is that this is exactly how the labyrinth forms as human embryos develop. The inner ears first appear as flat plaques of cells—placodes—one on each side of the developing head. Each placode then becomes concave, and then an increasingly deep pit. The edges of this pit then grow together and pinch off a hollow ball of skin and hair cells that then sinks deep into the head. Over the next few weeks, this hollow sac grows and distorts into an increasingly complex shape until it attains the configuration of your adult labyrinth. This is one of those haunting cases in which part of the body almost seems to replay its evolutionary history as it forms inside the developing embryo.

The first person to find his way around the labyrinth—the anatomical Theseus, as it were—was by all accounts an arrogant, sarcastic, domineering, rude, vindictive, artist-anatomist-genius named Antonio Scarpa. Born in the Veneto, he ruled the anatomical department of the University of Pavia (the university at which the almost-as-vituperative Golgi was to work a century later) with an iron fist for over fifty years at the end of the eighteenth century and the start of the nineteenth. A qualified doctor at eighteen years old, teaching university-level anatomy at twenty, the discoveries of this brilliant but flawed man would fill a shelf of books. He dissected, drew, and wrote on many aspects of the body, including the bones, the muscles of the thigh and groin, and the nerves of the heart. He studied the birth defect known as club foot and conducted fantastically delicate dissections of the inner ear, and it is for these that Scarpa is best remembered today. Many ear structures are named after him—for example the fluid inside the labyrinth is sometimes known as Scarpa's fluid. He occasionally toyed with surgery (and for someone like Scarpa that meant having two simultaneous professorships), but it was none other than Napoleon who insisted that he dedicate himself to anatomy. Perhaps it is not so inappropriate that after his death, Scarpa's head was cut off by his assistant and is displayed to this day in the university museum. Maybe this goes to show that you should be kinder to your subordinates than Scarpa was.

The internalized bubble of the inner ear labyrinth was used by our distant marine ancestors in exactly the same way that we use it now. The uppermost part of the labyrinth is distorted into three hollow tubes that arc out of the main body of the organ, each tracing part of a circle and then rejoining the rest of the labyrinth. Almost all vertebrates have three of these semicircular canals, arranged at neat right angles to each other—although for some reason the few surviving species of jawless fishes seem to cope with two, or even one. These canals tell us whether our head is turning or not, and most of us have three in each inner ear because we active, acrobatic vertebrates are free to tumble in three dimensions through our environment, whether it be the sea, air, trees, or grasslands. The end of each semicircular canal widens into an ampulla, or "flask," and within that widening lies a cluster of hair cells whose cilia lie embedded in a little, gelatinous lump called the cupula, or "little handle." This cupula sways freely in the liquid sloshing through the canals, and this is the key to how the whole apparatus works. As your head turns in one direction, the labyrinth also moves in that direction. However, at first the fluid inside the canals tries to stay where it is while the labyrinth moves around it—think of what happens when you try to spin a bucket of water. The cupula is thus tugged to one side by the fluid flowing past, and this tweaks the sensitive hairs embedded in its base. Remarkably, the hair cells from which these hairs protrude convert this tweaking into a signal that is then conveyed to adjacent nerve endings. Remember that Cajal showed us that these nerve endings do not touch the receptor cells, so the information must jump across the gap somehow. These nerve cells have their cell bodies in Scarpa's ganglion, right next to the labyrinth, and their axons then travel the length of the vestibulocochlear nerve back to the brain.

Perhaps more startling, the impulses from these tweaked hairs can actually be used for something useful. In fact, much of the hindbrain is devoted to interpreting this turning information. Not only are you consciously aware of turning in space, but your hindbrain also triggers several subconscious reflexes in response to that turning. For example, when your head turns one way, your eyeballs turn in the opposite direction so that you can maintain a fixed gaze—a mechanism called nystagmus, or "nodding." Also, if your labyrinths sense that you are falling over, your hindbrain automatically takes steps to protect you. If you start to fall to your right, commands are immediately sent down your spinal cord. These straighten your right leg in an attempt to correct the lean and also straighten your right arm to break the fall. This is quite a difficult reflex to suppress consciously— try falling sideways onto a mattress and see if you can keep your limbs relaxed until you land. So strong is this compensating reflex that if animals suffer brain damage which disconnects the hindbrain from higher regions, the balance centers take over and thrust the limbs into a disturbingly uncontrolled straightening called extensor rigidity.

There are some interesting side-effects of detecting spin by fluid sloshing about in our inner ear canals. One is that when the head turns, the fluid in the labyrinth does not stay static forever. Within seconds the fluid in the canals starts to be dragged around by the labyrinth walls and soon both fluid and walls are turning at the same rate. This explains why you feel dizzy when you jump off a merry-go-round. As your feet hit solid ground your labyrinths stop spinning, but the fluid within them continues to rotate for a few seconds, making you feel unpleasantly dizzy as well as inducing a flicking of the eyeballs—"post-rotational nystagmus." You can also induce dizziness, eyeball-flicking, and nausea by pouring ice water into one ear for thirty seconds or so and then raising your head. One of the three semicircular canals lies nearer the surface than the others, and the chilling effect of the water causes eddies of fluid within this canal, giving a false sensation of turning. The effect is quite dramatic and has caused people who were not holding on to some immovable object to fall over and fracture their skulls.

One of the most familiar side-effects of our semicircular canals is the feeling of the room spinning when you lay your head on the pillow after consuming one drink too many. One of the reasons that the semicircular canals work well is that the cupulae are the same density as the fluid within which they hang. If you drink large amounts of alcohol, some of that alcohol makes its way into the fluid inside the labyrinth, making it less dense. The cupulae now sag under their own weight within the lightened fluid and tug on the little hair cells. Your wonderfully tuned spin detection system is now mistakenly detecting gravity, and so you feel as if you are constantly turning. Still worse, if you lie in bed and close your eyes, you extinguish the one main line of evidence which reminds your brain that the room is not, in fact, spinning—the evidence of your eyes. Of all the things I teach my students, for some reason this is the one nugget that they all remember.

Although it is noticeable how little the semicircular canals have changed since those ancient ancestors of ours first writhed through the oceans, the lower part of the labyrinth has changed and diversified a great deal. However, its initial configuration was simpler than that of the semicircular canals—just a few hollow swellings. We do not really know how many were present, but there are many vertebrates that now have three swellings in the lower labyrinth—the sacculus, utriculus, and lagena; or "little bag," "leather pouch," and "flagon." We humans only have the first two (see Figure 8.2). As we will see, strange things happened to our flagon.

Knowing whether it was spinning or not was simply not enough for our ancient vertebrate ancestor—it also needed to know whether it was moving forward, sideways, or up and down, and the sacculus and utriculus allowed him to do this. These widenings of the labyrinth each contain a structure not dissimilar to the cupulae in the semicircular canals—little gelatinous lumps with the tips of hair from hair cells embedded in their base. These lumps, called maculae or "spots," are slightly different from the cupulae, and this turns out to be crucial to their role. In some sharks and rays, particles of fine sand are sucked down the endolymphatic duct and become embedded in the maculae, making them denser than the surrounding Scarpa's fluid. Unlike sharks, we do not go around sucking sand into our inner ears, so we have to make our own—our maculae are peppered with tiny crystals of dense calcium salts called otoliths ("ear stones").

Many animals form these otoliths throughout their lives, and ichthyologists can even age fish by slicing through otoliths and counting the rings, just like the rings in a tree trunk.

The density of the otoliths means that when the head starts to move in one direction, the heavy maculae lag behind slightly and briefly tweak the hairs of the underlying hair cells. Once again, these hair cells convert that tweak into nerve impulses in the neurons of Scarpa's ganglion. These impulses are then conveyed into the brain in the same vestibulocochlear nerve that connected to the semicircular canals, and the brain then uses the impulses to calculate the acceleration of the head. Each macula can potentially detect acceleration in two dimensions, so two maculae are more than enough for our three-dimensional world. An additional advantage of having little accelerometers in your head is that they can also detect the force of gravity.

Evolutionary biologists are not sure when, but at some point in evolution one of the maculae started to serve a different end—detecting the pressure waves of low-frequency sound. This is more difficult to detect than acceleration, but with a little cellular adjustment the inner ear was able to extend its repertoire. In fact, it is quite likely that your maculae can still detect some low-frequency sounds, but there is now a new part of your ear that does the job much better. Although sound travels very efficiently through water, we do not actually know how well a fish can hear with its lagena, the lowest swelling of its labyrinth. Some probably hear better than others, and some fish with air-filled swim bladders have coupled them up to their inner ear and use them as acoustic detectors. Nature is good at adapting what is available, and some knobbles of bone fall off the frontmost vertebrae of developing carp and catfish and form a bony acoustic linkage between swim bladder and labyrinth. Named after the comparative anatomist Ernst Heinrich Weber, these Weberian bones allow these species to hear higher-frequency sounds than the dull thuds and throbs detected by other fish.

Around 400 million years ago, one fishy lineage hauled itself up onto the land and was innocently to give rise to all land vertebrates. Although their senses of spin, acceleration, and gravity worked fine on land, the early amphibians faced some real acoustic challenges. They probably could not hear very well anyway, but now they became almost deaf because sound waves do not transfer well from air to a liquid-filled inner ear. Different groups of early land vertebrates opted for different solutions to this hearing impairment. Some started to concentrate on detecting sound vibrations from the ground—for example by adding a conducting strut of bone between their forelimb bones and their labyrinth. Although we humans cannot detect sound coming through the ground, it is entirely possible that many animals can, and this may even explain the apparent ability of some to predict earthquakes. A very clear use of this seismic ability occurs in certain lizards that rest their lower jaw on the ground to convey sound waves to their inner ears—not unlike the way Ludwig von Beethoven, profoundly deaf, wedged a stick between his piano and his skull as he composed.

Another solution to the problem of hearing in air is simply to make the labyrinth bigger. One of the strangest attempts at this strategy occurs in frogs, in which the two endolymphatic ducts can extend along each side of the spinal cord all the way to the end of the body, presumably turning the back into a large, slimy green microphone. It has even been suggested that the famous sail-backed Dimetrodon, the mammal-like reptile so familiar to dinosaur-obsessed little children, used a similar mechanism, and that its ungainly sail was actually a huge acoustic detector coupled up to grossly enlarged endolymphatic ducts. We would love to know more about this because Dimetrodons are more closely related to us than to dinosaurs. To what were they listening so intently?

There is a rather less ungainly solution to the hearing-in-air problem. It is such a neat one that it appears to have evolved several times in different land vertebrates—at least five times: in frogs, turtles, lizards, mammals, and dinosaurs and the birds to which they gave rise. The solution is to make an acoustic coupling that efficiently converts sound waves in air into sound waves in Scarpa's fluid, and it requires the formation of a new air-filled cavity just to the side of the labyrinth. Fortunately, early land vertebrates happened to have a handy cavity they were not using much anymore—the first gill. These creatures formed a pair of outpouchings from the sides of their throat, but instead of punching them through to the outside world to make a gill, they retained a thin, membranous barrier between the cavity and the outside. You still have remnants of these structures on each side of your head—the outpouching is your middle ear and the membrane is your eardrum. This is why your middle ear connects up to your throat via Eustachian tubes. This connection has its advantages and disadvantages: by swallowing you can relieve the pressure that builds in your ears as your plane lands, but throat infections can spread to give you an earache.

A middle ear cavity is not in itself much of an innovation. But the stapes bone, which spans it, conducts sound from the eardrum to the labyrinth. The stapes is the smallest bone in your entire body and has a distinctive stirrup shape, which gives it its name. It evolved from a chunkier predecessor that holds most of the back of the skull together in fish and probably controlled water flow through the first gill of some of our ancestors. Anyway, in reptiles and birds, one end of the modern-day stirrup is stuck to the eardrum and the other is stuck even more firmly to a tiny membrane on the surface of the labyrinth, the fenestra ovalis or "oval window." The clever thing about the whole system is that the eardrum is bigger than the oval window, and this is why sound hitting the eardrum is transformed so efficiently into sound energy in Scarpa's fluid. (You can either take my word for that, or if you have a mathematical bent, consider that the ratio between the areas of the eardrum and oval window is equal to the square root of the ratio between the acoustic impedances of air and Scarpa's fluid.)

See—I told you hearing was clever. That sort of physics does not make for an exciting read, but it does show just how elegant and resourceful evolution can be when it has a difficult problem to overcome. I think it is strangely gratifying that tortoises' ears are doing something that takes university-level physics to understand.

Soon an auditory arms race was underway—predator and prey were developing ever more sensitive hearing systems to gain the upper hand. Every new innovation was matched by a counter-innovation. The sound conduction mechanism of the middle ear was now so efficient that it showed up the shortcomings of the lagena of the inner ear. It was rather like recording a symphony orchestra on a cheap plastic cassette recorder with a pillow on top of it. Vertebrates solved this problem by developing an improved biological microphone with which fully terrestrial reptiles, birds, and mammals have been tinkering right up to the present day. It is called the organ of Corti, after the nineteenth-century anatomist Alfonso Corti (another alumnus of the University of Pavia).

The organ of Corti is an altogether more delicate mechanism than the lagena. Instead of being jammed into a sticky macula frosted with heavy otoliths, the hair cells in the organ of Corti are delicately inserted into the thin tectorial ("roof") membrane. Being so small, this membrane allows the whole apparatus to vibrate more easily and at higher frequencies, making its owner able to hear much quieter sounds and much higher-pitched sounds. The old labyrinth was not large enough to house this innovation, and so it added an annex, which is called the cochlea.

The mammalian cochlea is a long, thin structure that had to be coiled into the shape of a snail shell to fit into our little ancestors' skulls—cochlea is Latin for "snail shell" (Figure 8.3). The coiling may not actually serve any acoustic purpose, but if this is true, then it is difficult to explain why the mammal with the most cochlear coils is the largest rodent in existence, the capybara. Anyway, the coiling has been unwound in the illustration. When unwound, the two-and-a-half turns of the human cochlea is only three centimeters long. All along its length the cochlear duct is sandwiched between two other fluid-filled tubes. The fluid those two tubes contain is very different from the Scarpa's fluid inside the cochlear duct. These tubes are called the scalae, or "staircases," because their helical configuration looks like a spiral staircase.

The sound vibrations from the wiggling stapes are transferred into the upper scala through the oval window. They then pass along it right to the tip of the cochlea, where they escape through a tiny hole called Scarpa's hiatus, or the helicotrema ("hole in the helix"), into the lower scala. They then travel back along this to the round (or Scarpa's) window, whose wiggle then reciprocates that of the oval window. The net effect of this surfeit of Scarpas is that the cochlear duct vibrates between the scalae, and the organ of Corti within it thrums in sonic sympathy. And the thirty-two thousand hair cells in each ear are gently plucked.

But to me the truly wondrous thing about the long, thin mamma-

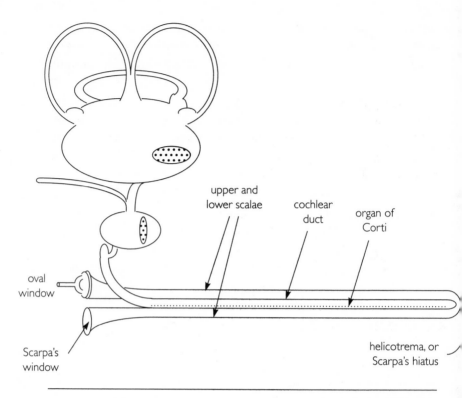

upper and
lower scalae   cochlear
duct   organ of
Corti

oval
window

Scarpa's
window

helicotrema, or
Scarpa's hiatus

Figure 8.3 The lower part of the mammalian labyrinth is made up of a coiled, blind-ending tube called the cochlear duct. The coiling is simply to pack the tube into a small space, and not for any acoustic purpose. Unwound, as pictured here, the human cochlea is roughly three centimeters long. The cochlear duct is sandwiched between the scalae—two other fluid-filled tubes that connect at the tip of the cochlea through a tiny hole called the helicotrema. The other end of the upper scala ends at the oval window, where the stapes imparts sound energy into it. The lower scala ends at the round, or Scarpa's window. The actual sound detector is the long, thin organ of Corti, marked with a dashed line.

lian cochlea is that it is the first stage in our ability to discriminate the pitch of sounds. Without the cochlea we would not hear music. The cochlea is not the same all the way along—in fact it changes continually and predictably from base to apex. The membrane on which the organ of Corti sits is thin and rigid at one end and thick and pliant at the other, and as a result, sound waves of different frequencies are muffled at different rates as they propagate along it. The end of the organ of Corti near the oval window and stirrup vibrates at high frequencies and the end near the helicotrema vibrates at low frequencies. And as the nerve fibers exiting the spiral ganglion at the core of the cochlea are bundled up in the order they connected to the organ of Corti itself, the brain receives nerve fibers arranged in a tonal "map" of the sound the ear is receiving. The cochlea is often compared to a resonating organ pipe, but really it is far more sophisticated than that—more like a microphone connected to a digital sound frequency analyzer. All this in just over an inch of tiny, membranous tube.

The mammalian ear has evolved two even more useful innovations that improved its ability to serve as a microphone. The first is that it has become an active system—the brain not only receives sensory information from the cochlea, but it is also constantly modifying the cochlea's responses to sound. We now think that only a quarter of all the hair cells in your cochlea are sensory—the other three quarters are being constantly driven to vibrate by motor impulses from the brain. By wiggling selected regions of the cochlear membranes, these active hair cells can suppress spurious oscillations that reduce the ear's ability to discriminate tones—they reduce the "blur" in the sound signal. This allows us to determine pitch much more precisely, and possibly helps us distinguish among several sounds all heard at the same time. This active control of the cochlea is why Ramón y Cajal did not use the ear to show the direction that impulses pass through neurons—most of the nerve connections to the ear are motor, not sensory, with dendrites pointing to the brain and axons pointing to the sense organ. It also explains one of the weirdest phenomena in neurology: otoacoustic emissions. Because most of the hair cells are forcing the cochlear membranes to vibrate, the cochlea actually produces sounds, and sometimes these sounds are audible to an outside listener. For example, pediatricians have reported tones being emitted from their patients' ears. This is not the same thing as tinnitus, or "ringing in the ears," in which you yourself hear the tones—these otoacoustic emissions are genuine sounds chiming out from normal ears.

The other great mammalian achievement came rather unexpectedly when two other ear bones were inserted between the eardrum and the stirrup. These bones, known as the malleus ("hammer") and the incus ("anvil"), originally formed the joint between our jaws and our skulls, but somehow in the course of evolution they were snapped off and squeezed into our middle ear (and two different bones were recruited into forming the jaw joint). The addition of two more bones in the chain did not really improve sound conduction from air to inner ear, but it allowed us to control that conduction. You have probably noticed that if you hear an extremely loud noise, such as a gunshot, your hearing is muffled for a while afterward. This is actually a protective reflex in which a tiny muscle pulls on the hammer to tauten the eardrum, damping its vibrations and thus protecting the sensitive cochlea. Although this is a crude protective mechanism, there is more subtle control in the movements of the stirrup. The smallest named muscle in the body, the stapedius, tugs on the tiny stapes bone to modulate the sounds entering the cochlea, and to remarkable effect. First of all, the stapedius tugs on the stirrup when we ourselves vocalize, so we do not deafen our sensitive mammalian cochleas with our own voices. Second, it damps out low frequencies, greatly increasing our ability to hear high pitches. We now think that it is mainly by the sterling efforts of this minuscule muscle that we mammals can hear much higher tones than other vertebrates. It has even been suggested that this is why infant mammals make high-pitched squeaks to communicate with their parents. Were these high frequencies a "safe channel" for communication in an era when dinosaur and avian predators were lurking just outside the burrow? Think on that the next time Junior is wailing.

The convergence of acoustic innovations in early mammals was extraordinary. It was as if these tiny animals were listening to compact discs while everyone else was still listening to wax cylinders.

Desert mammals are a good example of the amazing feats of hearing that mammals can achieve—probably because they cannot always smell much in the baking sun and often cannot see over the next dune. In contrast, sounds travels well in the clear air and over treeless sands. Desert mammals usually have huge ear flaps—another mammalian innovation—to funnel sound waves in, and they can often swivel them in different directions to pinpoint the precise origin of each and every sound. Domestic cats are probably descended from desert-dwelling cats, and this explains the two radar dishes they lug around on their heads. Desert animals also frequently have enormous middle-ear cavities, delicately sculpted to enhance sensitivity to high frequencies even further—some are so large that the right and left cavities touch each other at both the top and the bottom of the skull. Cats' middle ears are quite modest in comparison with some, but they contain a delicate, bony baffle that increases their ability to hear both high- and low-frequency sounds.

We often think of our hearing as being somewhat inferior to that of other mammals. Whatever feature of hearing you look at, there will always be an animal better at it—high frequencies, low frequencies, extremely quiet sounds. In comparison we humans seem rather mediocre—adequate, but not exceptional. I would argue that this is not always the case and that in some ways we excel. I wonder if it is because we are so visually oriented that we appreciate our hearing abilities less. One thing we are very good at, for example, is picking out individual sounds in a noisy environment. Of course lots of animals can do this, but our ability to follow the erratic fluctuations and cadence of an individual voice in a crowded room of ostensibly similar voices is really nothing short of remarkable. We are arguably the most verbally social animal, and so perhaps it is no surprise that picking out a lone voice among the chatter is exactly the sort of thing we are good at.

There are other aspects of hearing that we are only beginning to understand. One thing that many people report, especially when they are deprived of the sense of vision for some reason, is that they can somehow sense large structures such as walls and open windows around them. They often report that they can "feel" a very slight pressure pushing back on their faces from nearby objects. Blind people report this most often, but experiments have shown that it is a skill that many of us could hone if only we needed to. For a while, scientists thought that this sensing of objects was something to do with detecting airflow over the face, perhaps using facial hairs like animals use their whiskers. However, the ability to sense objects in this way is largely extinguished by wearing earplugs, and so it seems that this is in fact an unusually refined form of hearing. Our everyday world is never silent, partly because we ourselves are constantly flooding our environment with the sound of breathing and footsteps. It may seem unlikely that we could interpret echoes of this sound from nearby objects as some sort of acoustic "confinement," but this is probably exactly what we do. We are much better at picking up these subtle acoustic cues than we think—we are often surprised how different a room in our home sounds when we remove carpets and soft furnishings to do some decoration. Most often visually dependent, we tend to underestimate our hearing. Also, the way that we interpret sound echoes as pressure on the face is an interesting sign of how we can subconsciously merge different senses—a merging called synesthesia, to which we will return in Chapter 14.

Perhaps, then, we are not so very different from the animals that we admire for being able to echolocate. Bats and cetaceans are really just doing what we can do, but a great deal better. A common feature of these two most modified groups of mammals is that they both emit high-frequency sound, ultrasound, from their upper respiratory passages and then wait for echoes to be reflected back from the objects around them. The advantage of ultrasound is that it allows them to detect smaller objects—bats, after all, have to detect insects as small as midges. As ever higher frequencies are used, however, their ability to penetrate both air and water wanes, and so both groups actually emit a range of frequencies to locate both small and distant targets. For example, many bats send out frequency "sweeps"—whoops that start at a low pitch and slide up to a higher pitch. Cetaceans often send out a short burst in which all these frequencies are mixed in a single chord, but although this takes less time, the signal-processing challenge facing the brain when the superimposed echoes of these tones are reflected back must be enormous.

Echolocation also places tremendous demands on the sensitivity and resilience of the organ of Corti. Sounds emitted by echolocating mammals can be extremely loud—dolphins can focus theirs into narrow beams that stun their fishy prey and even marauding sharks. They achieve this focusing with a gelatinous, acoustic "lens" called a melon, a large, globular shape that lies above the skull. This is the dome shape on the top of dolphins' heads that makes them look so intellectual. When echolocating, bats and cetaceans are faced with a problem: the reflected sounds they wish to detect are far, far quieter than the bursts of noise they must emit. Thus they must protect their inner ears from destruction by the intense sound energy they are generating within their own heads. In bats at least, we think this is achieved by the furious activity of those tiny, ear-protecting muscles in the inner ear. As the whoop is emitted, these muscles immobilize the middle ear bones, but they must then relax before the incoming echoes arrive. Even with this protective mechanism, it is still hard to see how the organ of Corti is not shattered by sound waves traveling directly through the skull.

Those little neuromasts have certainly come a long way since they first lined the flanks of those early proto-fish. Again and again we have evolved new ways to use them—as the lateral line of fishes, to detect balance, gravity, and even sound. Those ciliated cells will appear again later when we investigate the other senses, but we will look next at where all their movement and sound information is going. As we will see, etched compellingly in the geography of the brain is evidence that the senses of hearing and balance were the driving force in the evolution of the hindbrain, the next area in our journey through the central nervous system.

9

THE BRAIN AS
ARCHAEOLOGY

The Hindbrain

Why are there three bulges in the brain? Why hindbrain, midbrain, forebrain? Scientists have wondered about this for a couple of centuries now, but no one has come up with a complete explanation. We would usually be happy to accept this sort of thing as an accident of evolution, but the fact that all vertebrates, no matter how weird and wonderful, have a three-bulge brain really invites speculation. What could explain this holy trinity of neurology?

Perhaps the best thing to do is to think back to why the brain is there in the first place. Animals do not have to form a large concentration of nerves in one part of the body, but many of them do. As it so happens, these concentrations are almost always near the mouth, or near major sense organs, or both. We can easily accept that our brain formed to be in convenient proximity to the mouth or senses, but it is difficult to explain how the mouth might then make the brain subdivide into three sections. There is, in short, nothing very tripartite about the mouth. The special senses, however, are a different matter. For some time, biologists have noticed that the three special senses located in the vertebrate head—smell, vision, and hearing—connect neatly to the three different brain bulges—the forebrain, midbrain, and hindbrain, respectively. We no longer think that this is just a coincidence—we now believe that the inputs of these three senses were the impetus behind the evolution of the three bulges.

This is the main reason why the central portion of this book is about senses—they explain the structure of your brain. Although sense organs are outside the brain itself, they have dominated it throughout its existence. After all, what is a brain without any information coming in? I admit that there are complications to this theory, but they can often be explained by deviations from the original vertebrate plan that have occurred relatively recently. For example, one problem is that in humans each of the three brain bulges does not correspond exclusively to its own sense. The supposedly smell-dominated bulge also processes visual and sound information, for example. This is not too much of a conceptual hurdle, however, as it simply reflects how brain regions can take on new roles over the course of evolution. This may sound like a weak response, but we can actually see this cerebral "mission creep" occurring all the time in different vertebrate groups. The fact that we humans have shifted more and more sensory processing into our forebrain is simply the way we decided to do things. We just seem to like mixing our senses.

Anyway, bear with me and I hope that, with a few admitted flaws, the idea of the three great senses giving rise to the three great brain regions will convince you. We will start with the hindbrain—a tangle of hearing, balance, and a little bit more.

Your spinal cord merges into the lowest part of the hindbrain as it protrudes slightly from the foramen magnum, or "big hole," at the base of your skull. In less respectful times, all the brain and spinal cord was called the medulla, which simply means "marrow"—the soft stuff you find when you smash a hole in skull or spine. It is noteworthy that the Greeks and Romans independently hit upon the same dismissive term for these organs as the ancient Egyptians. The lower half of the hindbrain, claimed to be an amorphous, rectangular slab of brain tissue, was called the medulla oblongata, and this term has stuck to the present day. We have now forgotten that "medulla" was once the generic term for the whole central nervous system, so we often confusingly omit the "oblongata" bit of the name. We toyed with the more exciting-sounding "epipsyche" for a while but soon tired of it. So "medulla" it is, but amorphous slab it is not.

As we move higher, the medulla gradually widens. You can see this gradual widening in Figure 4.7, a slice of my own brain. Later in the book I will tell you why this scan was carried out. The medulla is not simply a chubbier version of the spinal cord. You may recall that the extreme kinking of the embryonic hindbrain changed its configuration forever. Just like the over-flexed banana of my over-stretched analogy in Chapter 4, the back of the hindbrain has almost split open. As we edge up from the cord to the medulla (Figure 9.1), the central spinal canal veers toward the back surface of the neural tube. When it reaches that surface, the fluid of the spinal canal is separated from the outside tissues by no more than a thin, translucent membrane. The point where that membrane first appears on the back of the central nervous system is where we arbitrarily locate the junction between cord and medulla. It is marked by a little bulge of cord tissue with perhaps the most Scrabble-friendly name in all neuroanatomy, the obex, which means "barricade." Above the obex the fluid-filled ventricular space progressively widens and flattens to form the diamond-shaped fourth ventricle. The ventricle continues to widen throughout the medulla, but once in the top part of the hindbrain, the pontine region, it narrows down once more to form the top point of the diamond.

So the fourth ventricle is held in place by a thin covering, rather like the covers that stop leaves falling into swimming pools. This cover is called the velum, which means "sail-cloth," or perhaps more precisely, "awning." There are two portions of this awning—a lower one, pierced by Magendie and Luschka's holes through which the cerebrospinal fluid leaks from the brain, and an upper one, sometimes rather mysteriously called the valve of Vieussens. Thus the fragile, apparently inconsequential velum is actually a veritable biological hall of fame—François Magendie worked out the differences between the front and back nerve roots of the spinal cord as well as discovering the causes of severe allergic reactions; Hubert von Luschka has also given his name to ducts, joints, glands, muscles, and tonsils; Raymond Vieussens obviously had something of a valve fixation because he gave his name to yet another valve—a real one this time—in the veins of the heart.

In a brain that has been removed from the skull, the velum is so thin that you can peer straight through it at the floor of the diamond-lake of the fourth ventricle. The bed of that lake is not flat. Four subaquatic ridges run along its entire length from bottom to top, two on the left and two on the right. These ridges are not simply to add a

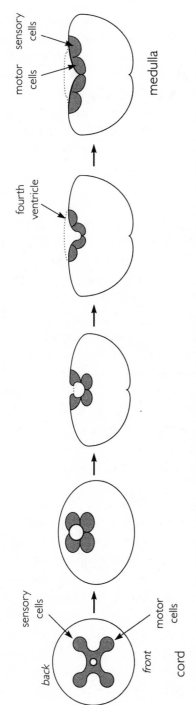

Figure 9.1. A series of schematic cross sections (see Figure 9.3 as well) showing how the spinal cord merges into the hindbrain. As the cord approaches the skull, its central canal lies farther and farther back until only a thin membrane separates it from the outside (*center*). The continuation of the canal into the medulla of the brain becomes broader and flatter—the fourth ventricle. As the fluid space in the center of the cord shifts, the motor and sensory cells shift with it. The motor cells at the front of the cord end up near the middle of the medulla, whereas the sensory cells at the back of the cord end up nearer the outside of the medulla.

little topographical relief to the lakebed—they are visual evidence that the layout of the cord extends right up here. As the central canal of the cord opens into the fourth ventricle, it drags the front and back wings of the spinal grey-matter butterfly with it (see Figure 9.1). These columns of motor and sensory neurons now end up under the lakebed of the fourth ventricle, where they form the four ridges—the two middle ridges are the motor cells and the two outer ridges are the sensory ones. So the pattern of the spinal cord is continued into the hindbrain, although in a distorted shape—as if some horrid little boy has torn the wings off the butterfly.

And just as the motor and sensory cells of the cord are connected to the body by spinal nerves, the cells in those lakebed ridges are connected to the head by their own series of nerves, the cranial nerves. They are a more varied group than the spinal nerves, mainly because the head has more varied components than the body, but they follow many of the same rules. Early anatomists discovered what looked like twelve pairs (left and right) of nerves emerging from the brain, and they gave them descriptive names as well as Roman numerals, from number I attaching at the front of the brain to number XII at the back, nearest the spinal cord. Unfortunately, the system has fallen into disarray with the more recent discovery that II and possibly I and VIII are not really nerves at all. Also, just to be irritating, it now seems that many nonhuman vertebrates have an extra nerve in front of I, but rather than renumber all the others, this has been allocated the distinctly un-Roman number 0. Snakes have no XI, but when you see what it does, this may not surprise you.

Anyway, if we draw a veil over these embarrassments, the cranial nerves still do the sorts of things that other nerves do. They send motor commands out to the head, and they bring sensory information back from it. As we sail upward through the fourth ventricle and explore the ridges in the depths below us, we can see the nuclei (Latin for "nuts"), the clumps of cells that are the source of seven of the twelve pairs of cranial nerves. We already know about nerve VIII:

XII. Hypoglossal, "under the tongue": moves the tongue
 XI. Accessory: moves some neck and shoulder muscles
  X. Vagus, "wanderer": movement of and sensation from most of the body's internal organs; taste sensation

IX. Glossopharyngeal, "tongue and throat": movement and sensation from the throat; taste and touch sensation from tongue
VIII. Vestibulocochlear: hearing and balance
VII. Facial: moves facial muscles; taste sensation
VI. Abducens, "leading away": moves some eye muscles
V. Trigeminal, "threefold": moves the jaw muscles; sensation from all face

Some of the lakebed bulges associated with these nerves are distinctive, others less so. For example, those joined to the large nerves XII and X can clearly be seen tapering down to a point at the obex, rather like the split nib of a pen—a formation called the calamus scriptorius, or "writer's nib." Some of the cranial nerve nuclei have disappointingly sensible names—the salivary nuclei of IX and VII control the salivary glands; the gustatory nuclei of X, IX, and VII deal with incoming taste sensations from the mouth. Some names are just plain unhelpful—the merged nucleus of XI, X, and IX that assertively controls many of the muscles of the throat, neck, and shoulders is cryptically named the nucleus ambiguus. Just one—the clump of sensory cells of X, IX, and VII—has an evocative name, the lonely-sounding nucleus of the solitary tract. You have probably noticed that one nerve seems to crop up again and again. Number X is called the vagus, or "wanderer," because it meanders all the way down your neck into your chest and abdomen and is your brain's link with almost all your internal organs. Remember this ambitious fellow, as we will meet him again.

All in all, the grey matter of the hindbrain is more patchy than in the spinal cord. There are gaps in the lakebed ridges. Some of the nerves have no sensory cells. Some have no motor cells. Some nuclei have merged together. And some nerves have cells scattered in several different nuclei. Clearly life is more unpredictable up here. Just to add to the challenges of medullary cartography, there are many splodges of grey matter outside those four lakebed ridges. In fact, most of the medulla and pons is a tangled mass of grey matter nuclei and white matter bundles. In many ways, the hindbrain is the part of the central nervous system with the least orderly structure, but this does not mean that it is a chaotic region. Its apparent disorder is meticulously reproduced in every newborn child, and so perhaps we should think

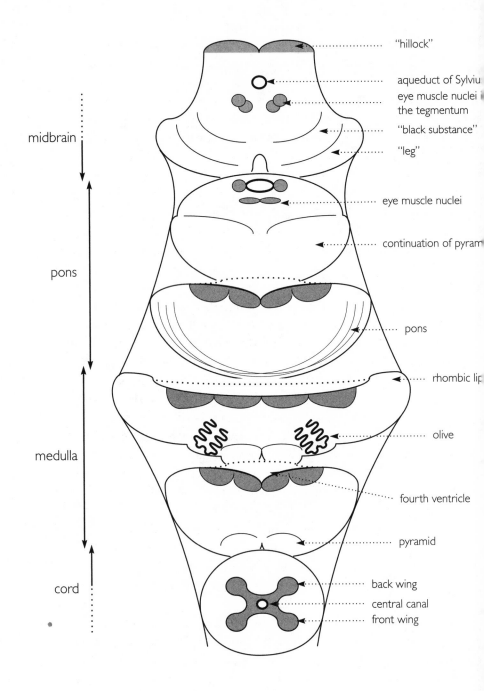

"hillock"

aqueduct of Sylviu[s]

eye muscle nuclei [of]
the tegmentum

"black substance"

"leg"

midbrain

eye muscle nuclei

continuation of pyram[id]

pons

pons

rhombic lip

olive

medulla

fourth ventricle

pyramid

back wing

central canal

front wing

cord

of it as a reflection of the varied and vital jobs the medulla and pons have acquired over the course of their evolution.

Much of the apparent disorder of the hindbrain results from the way that the cumbersome central processing centers of hearing and balance have to be crammed in. Of course it is appropriate that the medulla is full to the gunnels of cochlear and vestibular nuclei, busily processing the output of the inner ear labyrinths, because these centers are probably the reason that the hindbrain exists. As the roles of the labyrinth have changed over our evolutionary history, the nuclei hooked up to the vestibulocochlear nerve have simply been adapted to those new roles. Cochleas have replaced lateral lines, but the sound and motion processing of the hindbrain goes on.

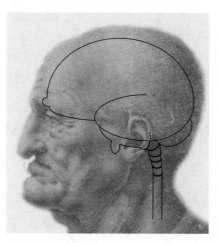

Figure 9.2. A series of six cross sections of the brain stem, from cord to midbrain. The front of the brain stem is at the bottom of each cross section. Above is a profile (by Leonardo da Vinci) overlain by an outline of the brain. The bold lines in the profile indicate the approximate position in the brain of each of the six cross sections in the main figure. Beginning at the bottom of the figure, the central canal of the spinal cord opens into a flat, diamond-shaped lake (fourth ventricle) in the medulla and pons, which then connects to the narrow aqueduct of Sylvius through the midbrain. The cerebellum has been omitted for clarity—it would be a large, globular structure atop the brain stem around the level of the pons and connected to it on each side by its peduncles.

The cochlear and vestibular nuclei are often so large that they are visible as distinct swellings on the sides or front of the brain. The cochlear nuclei also feed into huge nuclei near the front of the medulla, which are called olives because they appear as knobbles the size of small olives (Figure 9.2). The olives connect with many different parts of the brain, and their profuse axon outflow is bundled into characteristically crinkled white sheets called amicula or "cloaks," because they surround the olives like folded drapery. Even the way stations between the cochlear nuclei and the olives are large enough to protrude from the medulla—including the distinctively quadrilateral trapezoid bodies. All these gnarled protuberances that disfigure the human hindbrain give the impression that it is distinctly overstuffed.

The outputs of the hearing and balance centers are many and varied, and their outgoing axon tracts course up, down, and out of the brain. For example, balance may not be something we consciously think about very much, but it is the key to making coordinated movements. Because of this, the vestibular nuclei are wired up to all the major motor centers in the brain. They also have their own direct connection to the limb muscles via the spinal cord. It is this that makes you straighten the limbs on the side toward which you are falling. Most of the time, however, this vestibulospinal system just makes you stand or sit up, and this is why you do not have to consciously think about these complex activities, at least when sober.

Processing of sound information is scattered in a chain of nuclei extending all the way up the brain stem. Near the inflow of the vestibulocochlear nerve the sound information is arranged according to pitch, but soon various tiny sound-analyzing centers are hard at work. First of all, the tonal spectrum of the sound is sharpened. Helpfully, the organ of Corti often sends back pulses in synchrony with the incoming sound waves themselves, and the brain stem can calculate the exact frequency of incoming sound from what is thus effectively a digitized signal. However, this system of accurate pitch discrimination only works over a certain range—a range not dissimilar to that of a piano. In fact, we think this is why frequencies in this range sound musical, whereas very high and low notes sound inherently dissonant. Without the digitized sound signal we cannot be at all sure about pitch, so our brain seems to assign a jarring feel to high and low notes.

Perhaps the most important thing that animals need to know about sounds is where they come from. In many cases this can be a matter of life and death, so it comes as no surprise that much of the brain-stem processing of sound is dedicated to pinpointing its origin. The cells of the hearing nuclei achieve this by several mechanisms, some crude and some incredibly sophisticated. First of all, they work out whether the sound came from the left or right side of the head by dint of the simple fact that the head itself casts "shadows" in sound— sounds coming from your right sound slightly muffled in your left ear. The second feat of the brain stem is to work out whether sounds come from above or below, front or behind. This is a more sophisti-cated process. Your outer ear—the flappy bit on the side of your head—is completely asymmetrical. There is a lobe beneath, a curved scroll above, an abutment behind, and very little in front of your ear hole. Your brain can actually tell where sounds come from by the sub-tle alterations they undergo as they reflect off this varied aural topog-raphy. The third and most impressive stage is where the brain calcu-lates the exact angle from which the sound came. A sound coming from straight ahead reaches both ears at the same time, but a sound approaching from the side strikes one ear before the other. If you remember your school trigonometry, and someone told you the speed of sound, the width of your head, and the time lag between sound reaching the two ears, you could calculate the angle from which a sound has come. And this is what the brain stem does. If an inbuilt trigonometrical computer in your hindbrain were not impressive enough, this system also requires the brain to detect time lags as small as one-hundred-thousandth of a second. We still do not know how the brain manages such precise timekeeping.

So a chain of nuclei strung along the hindbrain is constantly pro-cessing and reprocessing the sounds we hear. By the time the infor-mation ascends to the midbrain, the cochlear, vestibular, trapezoid, and other nuclei have generated both a tonal map and a spatial map of those sounds. In humans the map can be so good that we can reach out in a pitch-dark room and seize the thing that was making the sound. In owls the map is so precise that they can swoop down on a scurrying sound source and grab it confidently along its body axis before carrying it off between the trees to its doom. And bats do the same with ultrasound. No wonder the hindbrain looks a little full.

All this acoustic equipment may distort the layout of the hindbrain by taking up a great deal of space, but it is not the only special sense crammed in these parts. Up to now, we have conveniently ignored the sense of taste, which feeds into the gustatory nucleus of the hindbrain through three pairs of cranial nerves. But was my argument not based on the idea that exactly one major sense organ in the head feeds information into one of the three brain regions, and that it was this sense that drove the evolution of that region? Surely we have a problem with taste and hearing both connecting to the hindbrain? Yet if we look closely at the history of the sense of taste, we soon find out that flavor is actually a mixture of at least four separate senses, some of which are spread all over the surface of the body in fish. To some extent, fish can "taste" the water in which they swim with every part of their skin. The fact that mammals have concentrated and internalized the sense of taste to the tongue should not blind us to the fact that we were once fish. Taste is not a sense rooted solely in the head, so I think we can exclude it from our scheme.

This is not to say that taste is not important. In humans, flavor has three or four different components. The first is the one with which we are most familiar—the specific detection of sweet, salty, sour, and bitter, as well as the more recently discovered "umami," a term for the taste of glutamate coined by a Japanese researcher in the early twentieth century. These tastes correspond to the binding of chemicals to ciliated cells in the taste buds of the tongue. Four of the five tastes are easy to explain—the cells simply bear specific molecules to detect sugars, sodium, acid, and glutamate. Bitter is rather more difficult to understand because a tremendously wide range of chemicals—often toxic ones—taste bitter to us. However, we now think that bitter-sensitive cells simply carry lots of different detector molecules. The second component of the flavor of a food is smell, and we will see later that this may be the most important of all. The third component is the feel of the food in the mouth—the melting of chocolate, the bristle of the kiwi fruit. And the fourth arm of our sense of taste is actually present to detect chemically induced pain. Fish have detector cells for noxious chemicals all over their body, but we have them in our mouth, and we now use them to detect hot chili peppers. These chemical pain-detector cells are more widespread than taste buds. We also have them in the eyes, anus, and vagina. This is the reason why extremely spicy meals can be as unexpectedly painful when they leave the body as when they enter it. It is also the reason why certain activities are probably best delayed after eating such a meal.

Taste is another sense that we often take for granted. This is probably because it very rarely fails, and those in whom it fails receive little sympathy. Yet taste is important for two simple reasons—it detects things we need like sugars and salt, and it warns us of plant toxins and unripe fruit. Of course modern human tastes have become somewhat depraved and many of us deliberately seek out bitter, acidic, and spicy foods, although it is noticeable that children are far more conservative about this. In many vertebrates, taste is even more important for sorting foods, and in these the central processing of taste can be extremely well developed. Many fish develop two huge gustatory bulges from their hindbrain that can exceed the entire forebrain in size. Sometimes these vagal lobes are so large that they contain their own outpouching of the ventricular system and dominate the whole brain.

Despite our best efforts, even if we discount the taste and hearing systems superimposed on this region, we still do not understand the construction plan of the hindbrain. For over two centuries, embryologists have tried to explain the hindbrain in terms of a segmented structure—almost as an extension of the spinal cord. Yet their well-intentioned attempts to extend this spinal segmentation further up into the brain have met with limited success. In addition, some nerves (V, VII, IX, X, XI) supply a set of structures that are segmental—the throat serrations which become gills in fish, and much of the lower half of the human head—but whose segmentation does not seem to bear any relation to that of the cord and so just seems to make life more complicated still. Potential salvation came with the late-twentieth-century discovery that the developing hindbrain has a regular series of internal bulges. These "rhombomeres" do correspond to the spinal and throaty segments of yore, but only in a very ad hoc manner—each segment receives nerves from a particular rhombomere, but the pattern seems arbitrary and incomplete. Some neurons even migrate from one rhombomere to another for no obvious reason. Has a once sensible and ordered system become confused by loss of ancient nerves, or has it been scribbled over by new structures acquired by messy evolutionary accumulation? Are we simply not intelligent enough to see the system behind the confusion, or was there never an orderly system at all? Maybe the hindbrain will make sense one day, or maybe it is the brain's insoluble Gordian knot.

The splodges of grey that smatter the hindbrain also include a number which coordinate many of our crucial visceral functions. For example, one need look no higher than the area postrema, the "very last bit" of the medulla by the obex, to find nerve cell bodies involved in vomiting—a life-saving protective mechanism in humans and animals. The area postrema is thus right next door to the calamus scriptorius, which contains the cell bodies of the vagus nerves to the stomach, the organ that must do the actual vomiting. Many cells that trigger vomiting do so when they "taste" noxious substances in the blood or even the cerebrospinal fluid. This internal "tasting" also extends to other brain functions. For example, other cells in your medulla detect acidity, which is a sign that you are accumulating carbon dioxide and need to breathe faster, and the nerve-cell circuits that coordinate, accelerate, and decelerate the rhythmic pattern of breathing are in, guess where, the pons and medulla.

The introspective nature of your hindbrain goes even further. It also monitors and controls your blood pressure and heart rate. Just as the thermostats and pressure valves attached to the pipes of a car feed signals back to a central computer, so the tiny pressure monitors and oxygen sensors feed information up the vagus nerve into the so-called cardiac and vasomotor centers of the hindbrain. These are little circuits of nerve cells that have the power to accelerate or slow the heart as well as constrict the body's blood vessels to raise blood pressure. Unlike the lungs, the heart has its own intrinsic rhythm and does not need nerves to help it beat, but the hindbrain is able to exert considerable influence over its speed, partly by means of the trusty vagus nerve. This nerve can exert a dramatic effect on the heart, even causing a fatal cardiac arrest. Unfortunately this can happen during, of all things, surgical manipulation of the eyeball—an inexplicable quirk of evolution called the oculocardiac reflex.

Weaving around and between all these clumps of grey matter are white matter tracts—bundles of axons carrying impulses around the place, and here again the hindbrain is rather a tangle. Obviously the neurons in the hindbrain have to send axons to each other, but they also have to communicate with cells in other parts of the brain and send axons down the cranial nerves, too. In addition, the hindbrain carries large tracts of fibers that are just passing through, *en route* between midbrain or forebrain and spinal cord. Really, the pons and medulla might best be thought of as the vibrant downtown of the brain, where little neuronal communities live and work in close proximity, squashed in by the axon traffic that weaves its way around them.

One of these axon tracts is so large that it is clearly visible from outside, coursing along the front side of the hindbrain. In humans it is hugely important: It is the main route by which the cerebral hemispheres send instructions down to the body to carry out conscious, manipulative movements. The name of this information superhighway changes a few times during its long course, but we will refer to it by the name it is given in the hindbrain, the pyramids (see Figure 9.2). So well developed is this tract in humans that it bulges forward as a pair of triangular eminences that could indeed serve as a backdrop for a train of tiny camels. The pyramids are far less prominent in most other mammalian species, even though they are in the same place. Horses, for example, are not known for making fine manipulative movements, and their pyramids fizzle out before they even reach the spinal cord.

There is a large sensory counterpart to the motor pyramids called the posterior columns, which bring sensory information from the body up to the cerebral hemispheres. Once again, their name changes as they chart their course. In the cord they were the tracts of Goll and Burdach, also known as the gracile ("slender") and cuneate ("wedge-shaped") tracts, but as they enter the hindbrain, they dive deep into its core as flattened cords to be renamed the medial lemnisci, or "middle ribbons." Once again, the ribbons are unusually large in humans. We will catch up with them later on, for they will be with us all the way to the uppermost parts of the brain.

In the medulla, the medial lemniscus and the pyramids both do something rather strange—they switch sides. The lemniscus from the left side of the body flips over to the right side of the medulla, and vice versa. And the downward-coursing pyramids also switch sides, most of them in the midbrain. The crossing over of the pyramids was first reported in 1709 by Domenico Mistichelli, a professor of medicine at Pisa. It must have seemed a complete mystery to him, much as it does to us today. This phenomenon is called decussation, which means "dividing by crossing over," and it is one of the reasons why the hindbrain is so packed with nerve fibers. Not only do axon fibers have to travel around, through, and beyond this region, but it also seems that for no obvious reason they must also cross from one side to the other as the mood takes them.

We do not know why this crossing over takes place, but it is a major feature of the brain that has an important effect on how we interact with the world. The general rule seems to be that fibers have to cross over if they are traveling to or from the forebrain. For reasons on which we will ruminate in Chapter 17, the forebrain is wired back to front—the left cerebral hemisphere senses and moves the right side of the body and the right cerebral hemisphere senses and moves the left side. We know of no good reason why having the forebrain wired in this way might help the brain work better, but it certainly creates some huge cabling problems lower down. The mid- and hindbrains are wired in a more straightforward way, as are the cord and the cerebellum, but that means that any connections between them and the hemispheres must cross over as well.

One of the largest fiber tracts in the whole central nervous system is a result of this strange tendency for crossing over. As we will see in later chapters, in mammals movements are mainly controlled by a constant interaction between the large superstructures tacked on to the brain stem—the cerebellum and the cerebral hemispheres. This means that there is a bustling two-way traffic of information between these two regions, but because the hemispheres are wired back to front, all that information has to cross from one side to the other. This decussation could take place anywhere, but evolution has decided that the best place for it is near the cerebellum (Figure 9.3). As I have already mentioned, the globular cerebellum grows out from the back of the hindbrain and is thenceforth held there on stalks, or "peduncles." In humans there are clearly three peduncles on each side, but this multiplicity of peduncles is less apparent in most other animals.

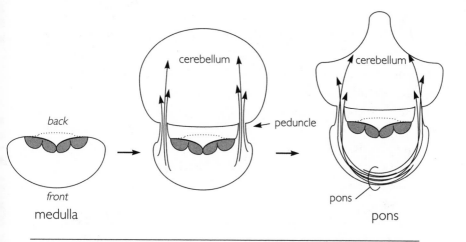

Figure 9.3. A continuation of the schematic cross sections of the cord and hindbrain begun in Figure 9.1, from the medulla to the pons. The ancestral upper hindbrain was probably similar in structure to the lower hindbrain—a flat slab of nervous tissue with the fourth ventricle behind it, covered by a thin membrane. In vertebrates, the cerebellum grows out of the side and front "lips" around the fourth ventricle. It is a large, globular structure that sits atop the hindbrain on peduncles, or "stalks." Nerve fibers pass to and from the cerebellum via these peduncles. In mammals, the cerebellum connects many times with the cortex. To allow this crossing from one side to the other, the pons, or "bridge" evolved to transfer fibers from the peduncles to the opposite side.

In nonmammals, most of the axon fibers in the peduncles travel to and from the brain stem or spinal cord, so they do not need to switch to the other side. As mammals evolved an ever-greater interplay between the cerebellum and the cerebrum, however, a thick band of fibers appeared, connecting the left and right cerebellar peduncles to the opposite side of the cerebrum. This band of fibers is called the pons Varolii, or "Varolio's bridge." Constanzo Varolio was the surgeon and professor of anatomy at Europe's oldest university, Bologna, in the late sixteenth century. Although he died at thirty-two years of age, his intellectual light burned brightly for that short time. He qualified as a doctor by twenty-four, was a professor two years later, and probably became personal physician to Pope Gregory XIII, who was also from Bologna. He made diverse contributions to anatomy, including rediscovering the muscles that stabilize the erect penis—Galen had found them, but his work had been lost in some prudish accident. Varolio is best remembered for inventing a systematic way of studying the brain by removing it from the skull and making repeated transverse cuts to observe it in cross section, rather like a modern MRI scanner. One of the structures he cut through on his meticulous travels was the bridge between the two sides of the brain stem that now bears his name. Unfortunately, we now rarely hear the eponym and the structure is usually just called the pons. The name is also used more loosely to describe the entire upper half of the hindbrain which it spans—it is simply a handier name than "metencephalon" or "metapsyche."

So the little hindbrain wedged onto the floor of your skull (see b and c in Figure 4.7) is overflowing with vital circuitry. Three different systems of head segmentation coexist with each other, bundles of fibers pass up, down, and from side to side, neural circuits maintain your vital functions and the signal processing of the senses of hearing, balance, and taste are squashed in, too. The hindbrain is a wonder of miniaturization that can seem confusing to the uninitiated. Yet the hindbrain is consistent—most functions are carried out in circumscribed little areas that are the same in everyone's brain, and because of this the hindbrain actually provides very rich pickings for neurologists trying to work out what has gone wrong in their patients. Because we know what all these structures do, the symptoms of patients with hindbrain disease usually make logical sense.

For example, damage to the nuclei of the cranial nerves can give a very clear indication of where brain injury has occurred. A glance at the list earlier in this chapter of the functions of these nerves is sufficient to explain why neurologists examine patients for signs of tongue paralysis, shoulder paralysis, difficulty breathing or swallowing, deafness or loss of balance, facial droopiness, abnormal eye position, and lack of touch sensation on their face, which correspond to damage from the bottom to the top of the hindbrain. Observation is often enough to determine which cranial nerves are damaged, but there is also a list of reflexes that specifically test certain nerves. For example, if you touch someone's eyelid, his nerve V announces this sensation to the hindbrain, which then orders nerve VII to shut the eyelids.

We can also glean useful information when large fiber bundles passing through the hindbrain are injured, just as we can in the spinal cord. But here we have an additional clue: We know where tracts cross from one side to the other on their way down from that wrong-way-round forebrain. A left-sided injury can cause either left- or right-sided paralysis, depending whether it strikes a major motor tract above or below the level where it decussates. Perhaps the most extreme and disturbing type of tract damage results from a complete destruction of all the tracts at the level of the upper pons. These patients lose all connections between their conscious, alert forebrain and almost all their cranial and spinal nerves. The only movements they can make are with their eyeballs, as most of the nerve fibers supplying the eyeball muscles exit the brain above the pons. Their eyes are now the only way they can communicate with the outside world—an eerie state known as "locked-in syndrome."

The most sensitive indicator of insidious hindbrain disease is often the pattern of breathing. We have seen that the circuits that maintain rhythmic breathing are in the hindbrain, as are the regions that speed it up and slow it down. Because breathing is controlled in this multicenter way, small areas of damage in many parts of the brain stem can cause subtle but distinctive symptoms. The first sign often comes when a patient is in dreamless sleep, when breathing would normally be at its most regular—and when there is no stimulation from a wakeful or dreaming brain to compensate for minor breathing irregularities. Breathing abnormalities can occur in almost any conceivable pattern, depending on their cause. One type is Biot's breathing, in which episodes of regular breathing are interspersed by alarming periods in which no breathing occurs at all. Cheyne-Stokes breathing seems quite similar, but the causes are probably different—periods of increasingly deep breathing are punctuated by phases of breath holding. One of the strangest conditions is respiratory dyspraxia, in which patients lose all conscious control of breathing, sometimes resulting in metronomically regular breathing. As you might expect, respiratory dyspraxia can make it very difficult to speak.

The most poetic breathing disorder is also one of the most frightening: Ondine's curse. Ondine was a water nymph in a French folk tale that has achieved the ultimate badge of tragedy—it has been adapted into an opera. According to the tale, water nymphs are not supposed to fall in love with mortal men, for bearing their children causes them to age like mortal women. Of course, this is exactly what Ondine did, and as soon as she bore a child to her beloved husband Palomon, her beauty began to fade. As she aged, the faithless Palomon gradually turned away from her and once again fell in love with his ex. Typical. One fateful day, Ondine found the two *in flagrante* and flew into a rage. Clearly not in the mood to negotiate an amicable separation, this woman scorned decided to act. Palomon had once promised to love her with every waking breath, so Ondine cast a spell on him that he would no longer remember to breathe when he was asleep. And remarkably, this is what sufferers of Ondine's curse do—they forget to breathe when asleep and must be artificially ventilated through a tracheostomy tube. Although the condition can be caused by injuries to the medulla, it often has a genetic cause. Strangely, though, it is not hereditary. It seems that victims inherit healthy versions of the culprit gene that are then damaged as they develop in the womb.

Before we leave the hindbrain behind us, we cannot ignore the fact that it is ground zero for some of the strangest and most destructive diseases in existence. Prion diseases begin silently, but slowly and unstoppably build into a crescendo of degeneration and devastation, ending in death. They can attack any brain cell, but they usually focus on neurons in our cherished hindbrain. We are pretty sure that we know what causes them, but it is not like the cause of any other kind of disease. They are infectious diseases, but you do not have to be infected to get them. There is an infectious agent, but it is not an invader. It is easier to get them if you are a cannibal, but you do not have to be.

Slow, fatal degenerative brain diseases have been recognized for some time. The first to be described was scrapie, which affects sheep, especially in Britain. Known for centuries, this disease has been sending sheep to their confused, staggering, untimely deaths for centuries without apparently posing much of a risk to anyone else, including unsuspecting Brits who have presumably been eating these sheep all those years. Scrapie occasionally spread to mink that ate infected sheep, but not many people were very worried about that. The name of the disease is a triumph of English plain-speaking—it was called scrapie because infected sheep feel incredibly itchy and compulsively scrape themselves on any nearby object. If people had been paying similarly close attention to wild deer, they would probably have realized that some of those get a similar chronic wasting disease.

Over the last couple of centuries, doctors started to notice similar conditions in humans, although they crop up in all sorts of situations. Creutzfeldt-Jakob disease is probably the best known and occurs most often when people are injected with extracts of other people's brains—such as when children are treated for growth-hormone deficiency with pituitary extracts. Two other human degenerative diseases, Gerstmann-Sträussler-Scheinker syndrome and familial fatal insomnia, added to the mystery, but it was the most exotic form of all that made headlines and sent us on our way to explaining these diseases. Kuru, also known as the "laughing death," infects women of the Fore tribe of New Guinea, who were known to honor their dead by ritual cannibalism. Kuru has now almost entirely disappeared, chiefly because the Fore have given up their unusual habits, but also because it made us realize that putting other people's brains inside your own body places you at risk. Human cannibalism is now probably quite rare. Not only has most ritual cannibalism largely died out, it is unclear if anyone ever ate other people simply for their nutritional content.

In the late twentieth century it became clear that the infectious agent for these slow diseases is very unusual. Unlike viruses, bacteria, and parasites, it contains no genes of any kind, so no one could understand how it propagated itself. Eventually it was realized that the agent was a protein, which was termed a prion—a contraction meaning "proteinaceous infectious thing." There is no known protein that can replicate itself, and so the hunt turned to finding out how the protein works. We are now aware that prions are actually distorted variants of normal proteins which our brains make all the time, but they have two features that make them deadly. First of all, they bind to the normal proteins in our brain and warp them into a similarly distorted form, thus rendering those infectious, too. Second, unlike the normal protein, neurons cannot dispose of the distorted diseased form. So if one prion gets into a brain, a runaway cascade of protein distortion is initiated that eventually clogs enough neurons to kill the victim. We know that distortion of normal proteins is the basis of these diseases because if the genes that make those proteins are "knocked out" of a strain of mice, they become invulnerable to infection. There is debate about how exactly the abnormal protein distorts the normal one, but that need not worry us here.

So some form of cannibalism is usually necessary for these diseases to spread, be that ritual, nutritional, or cannibalism by medical injection, or even by a bored sheep innocently chewing on its friend's recently discarded placenta. Yet there are ways in which the disease can start without any initial infection event. We think that, extremely rarely, a normal protein can spontaneously warp into the abnormal conformation. This is, to be honest, so uncommon that an individual person really need not worry about it. In a society that practices some form of cannibalism, however, this rare event can be the start of a storm of infections. And this is probably how all these diseases start. In addition, there seems to be another, hereditary form of these diseases in which people inherit genes that make protein variants which are especially prone to distortion—so prone that a distortion event becomes likely within a human lifetime. Because of this, these cases occur in family groups, completely unlike the infectious and spontaneous forms.

These rare noninfectious cases mean that these diseases will probably always be with us. No matter how careful we are about transferring tissue between people, there will still be the occasional unfortunates who are doomed by these diseases. We do not really know how prions kill people, but we are working on it. We know that they accumulate in slabs that jam the internal machinery of neurons, creating a characteristically spongy appearance—hence the name "spongiform encephalopathies." Yet many animals and people show severe symptoms when their hindbrain is not, to be honest, too spongy. The prions accumulate in structures we know well—the olives and the various motor nuclei of the hindbrain—but other diseases seem to do a lot more damage to these structures without causing such severe symptoms. Maybe the prions do not just clog up cells—maybe they provoke the immune system to do something disastrous, or maybe they damage lots of nearby nonspongy cells by causing the release of destructive chemicals.

One thing we would really like to know is what makes prions cross the species barrier. I remember listening to the radio on the day in the 1990s that the British government proclaimed that bovine spongiform encephalopathy, BSE, would probably not spread to people. At the time I recall thinking that this was rather a bold statement about a disease that had probably just spread from sheep to cattle. We also knew at that time that BSE had likely already infected a few cats. There was no evidence to say that it would spread to people, but no evidence to say that it would not. As it turns out, BSE hardly ever infects people, and this is of course good news for the beef-loving sector of the UK population. Nevertheless, BSE has certainly made us think again about how we conduct our farming. Dead sheep may have made economic sense as a protein source for cattle, but accountants are toying with powerful forces when they make such decisions.

10

BEAUTY IS IN THE EYE OF THE, ER, SQUID

The Origins of the Eye

People argue a lot about eyes. Especially creationists. Darwin started it when he wrote, "To suppose that the eye . . . could have been formed by natural selection, seems, I freely confess, absurd in the highest degree." He did not write this furtively in a private letter to a scientific colleague: instead he wrote it in a book called *On the Origin of Species*.

At first sight, if you will excuse the pun, the eye does look as if it has been carefully designed. The optics seem perfect and subtle—far more advanced than any artificial system. The light receptor cells are tiny and respond to light intensities over a bewildering range. Your gaze is steadied by an elegant system of cords and pulleys that continually adjust as the head and body moves or as your interest wanders to other things. Yet when we look back in vertebrate evolution, it seems as if the eye is as it ever was. Unlike the ear, with its well-paced history of modification and refinement, the eye just appears in the vertebrates as a *fait accompli*. Yes, there is variation among different species, but the basic wondrous design is there. We have no fossil record for the eye, so could it really have evolved by a series of undirected, accidental steps? If Darwin worried about it, then perhaps so should we.

What we do know about the eye is how it forms. Far more than any other sense, the eye is a product of the brain. Other tissues are in-volved, but without the instigation of the brain, eyes cannot form. The areas of the brain responsible are already decided at the neural plate stage, before the nervous system has even folded into a tube. Thus the eyes are in progress at an incredibly early stage. As soon as the brain has sealed into a tube, bulges start to protrude from each side. These little eye stalks are very much part of the brain, and they carry an extension of the ventricles within them and an outer coat-ing of meninges. As the eye stalks approach the skin on the outside of the embryo, they induce it to thicken into the optic placode, or eye plaque (Figure 10.1a). The eye plaque will form a crucial feature of the eye—the lens—but that structure probably evolved long after the eye stalks were already conveying visual information back to the brain.

The eye stalk and the eye plaque now commence an elegant series of twistings and foldings (Figure 10.1b,c). The tip of the eye stalk folds in on itself, becoming more and more concave until it takes on a shape not unlike a wine glass—a bulbous bowl connected to the brain by a thin stem. Because of how it forms, this optic wine glass is double layered. While all this is going on, the eye plaque on the surface of the head is itself becoming increasingly concave, forming a deep pit that eventually pinches off to yield a hollow ball of cells, much as the ear placode pinched off to form a bubble of hair cells. This hollow placode ball now hangs in the opening of the optic wine glass, rather like an olive skewered by a cocktail stick across its mouth—which is exactly where the lens needs to be. With that in place, cells in the general vicinity swarm around the optic wine glass and lens ball and form the outer walls of the eyeball. *Ecce oculus!* And behold—an eye.

There is a fundamental difference between the eye plaque and the plaques that form all the other sense organs—the eye plaque does not form any actual sensory cells. Instead it forms an optical structure, the lens. To do this, the back wall of the lens ball swells to fill its central cavity with neatly stacked living cells full of a transparent pro-tein called crystallin. Maybe this uniquely nonsensory fate of the eye plaque reflects the fact that it is an evolutionary latecomer—the lens may have been the last major part of the vertebrate eye to evolve. So

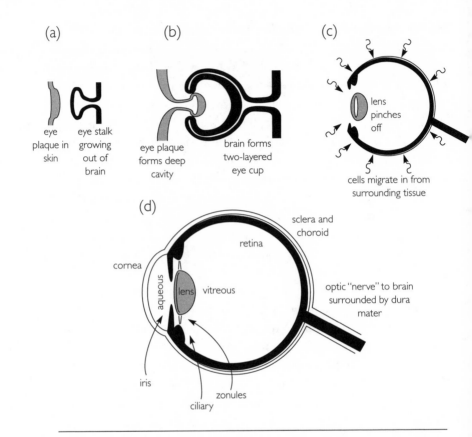

(a)

eye plaque in skin   eye stalk growing out of brain

(b)

eye plaque forms deep cavity   brain forms two-layered eye cup

(c)

lens pinches off

cells migrate in from surrounding tissue

(d)

sclera and choroid

retina

cornea

aqueous

lens   vitreous

optic "nerve" to brain surrounded by dura mater

iris

zonules

ciliary

Figure 10.1. The development of the eye. (a) The eye forms due to the interaction between two precursors. The first is a thickening in the surface skin of the head called the eye plaque (grey). The other is an outpouching of the brain called the eye cup (black). (b) The eye plaque (grey) becomes increasingly concave and folds away from the surface of the head. The tip of the eye cup (black) also folds inward to yield a two-layered wine-glass shape (black). (c) As the plaque and cup develop further to form the lens and retina, respectively, cells migrate toward the eye to form the other parts of the eyeball. (d) Much of the wall of the eyeball is formed by these immigrating tissues—the clear cornea, the white sclera, and the pigmented choroid (white). The lens is the only derivative of the eye plaque (grey). If the eye cup is considered to be shaped like a wine glass, the stem of the glass forms the optic "nerve" and most of the glass forms the two-layered retina. The rim of the glass forms two specialized structures. The front one is the iris—a pigmented, muscular ring that controls the size of the pupil. The back one is the muscular ciliary, which pulls on the lens via the zonules of Zinn, flattening it to focus on distant objects.

the lens is important, but it is not the part that turns light into nerve impulses. That job is done by that two-layered optic wine glass that sprouted from the brain. The bowl of the glass will form the retina, the sheet of light-sensitive cells, and the stem of the glass will form the optic "nerve," which carries the visual information back to the brain (Figure 10.1d). I say "back to" the brain, but as you can now appreciate, the retina is simply an extension of the brain, and so the optic "nerve" is really better called the optic "tract"—a bundle of fibers carrying information from one part of the brain to another. Not all of the retina lies where light can reach it. The front part of the retina, which is thinner, is called the pars caeca, or "blind bit." This is separated from the thicker pars optica by an irregular boundary beautifully named the ora serrata, or "jagged shore." As with the lens, I will say more about the retina after a further discussion about how our eyes evolved.

The rim of the optic cup has a different fate from the retina-forming part. The ring of tissue at its very edge thickens and starts to grow across the front of the lens, ending as a flat, circular ring with a hole in the middle. The hole is the pupil and the ring is the iris: the very thing that Santiago Ramón y Cajal filled with fragments from his cannon. So when you gaze lovingly into somebody's eyes, you are actually staring at the perforated frontmost extension of his or her brain, which I admit does not seem quite so romantic. Yes, the iris is brain—the window on the soul after all. Admittedly the iris is an unusual part of the brain. Beautiful pigmentation led to its name, which means "rainbow." Also, it forms its own intrinsic muscles to open and close the pupil, and so it is the only part of the brain that can move itself.

The ring of tissue just inside the rim of the wine glass also has an interesting and busy life ahead of it. Like the iris in front of it, this region becomes pigmented and muscular, although the actual muscle cells immigrate from outside the eye. It forms a delicate ring around the lens with fine radial serrations, each no thicker than an eyelash, that give the ring its name, the ciliary body (cilia is Latin for "eyelash"). These surface serrations are aligned with the tiny muscle fibers within the ciliary, so that when the ciliary contracts, it pulls outward on the edges of the lens and flattens it. The upshot of this is that light from distant objects is now focused onto the retina—an ability so astounding that it is difficult to imagine it evolving by random chance. The ciliary itself does not attach directly to the lens, however. Instead there are minuscule fibers around the circular edge of the lens that suspend it within the ring-shaped ciliary (see Figure 10.1d). These tiny tendrils, the most delicate part of all the brain, the most remote cerebral outpost, the gateway to the mind, are the zonules of Zinn.

Yet the ciliary has even more work to do. Not content with allowing us to focus, it also produces most of the transparent stuff inside the eyeball. As embryonic eyes grow, the surface of the ciliary slowly secretes a gel into the chamber behind the lens. This is the vitreous ("glassy") humor that makes up most of the mass of an adult's eyes. It is also one of the few instances in which the shunned, archaic word "humor" has been allowed to survive in modern medicine. Modern physiology long ago replaced the theory in which the health of the body was dependent on the amounts of four mysterious humors it contained, not unlike Galen's spirits (Galen thought that the eye and optic nerve were a route by which air could be sucked into the brain). You keep the same gel in your eyes throughout your life, and it is essential for keeping the retina stuck on the back of the eye. Inflammation can make the vitreous liquefy, causing the retina to detach and sag with disastrous effects on vision. The ciliary is also responsible for secreting the other aqueous or "watery" humor into the cavity in front of the lens. Unlike the vitreous, the aqueous humor is secreted and absorbed all your life. You probably get through several eyefuls of the stuff every day. Excessive pressure in the aqueous causes glaucoma, which can compress the retina into blindness if left untreated. A common cause of glaucoma is blockage of the outflow of aqueous humor at its usual exit, a structure that sounds like an Amsterdam suburb: the canal of Schlemm.

So the eye has its light-sensitive retina and its optically clear internal lenses of liquid, gel, and crystal. All it needs now is a rigid wall to bind it all together into a neat eyeball shape. The cells that swarm around the developing eye form a tough capsule for it—larger and white in the back and smaller and transparent in the front. As you may have guessed, the white part is the "white of the eyes" that one is supposed to wait to see before one shoots. Ophthalmologists call this the sclera, which is Greek for "hard." As if to emphasize the brainy affiliations of the eye, the sclera merges at the back into the dura mater, becoming more Latinate on the way, for "dura" is Latin for "hard." The transparent front part of the eye wall is the cornea, the bloodless, glassy dome through which we see the world. Apart from its lack of blood vessels, the structure of the cornea is remarkably similar to the sclera. It is tough, of course—cornea means "horny"—but by extremely careful control of its water content, it also manages to stay transparent.

There is one other component of the wall of the eye that is essential to its optical well-being. The choroid is a layer serving a variety of functions that is interposed between the outer fibrous sclera and the inner sensory retina. First, it is usually full of blood vessels and therefore responsible for bringing nutrients to the retina, which is often claimed to be the most energy-hungry tissue in the entire body. The second function of the choroid is purely optical—it is usually pigmented a soot-black color. If you unscrew the lens from an old-fashioned nondigital SLR camera, you will see that the inner workings are all black—even the little screws and springs. This is because a camera is meant to focus light into a clear image on film, so the last thing you want is for shiny surfaces to bounce light uncontrollably around the inside of the instrument and spatter the film with reflections. And this is exactly the same with the sooty choroid of the eye. Yet paradoxically, some animals' choroids have a third function, which is to deliberately cause reflections. We all know that a cat's eyes caught in the headlights reflect light back at us. This light is bouncing back from a specialized layer of the choroid called the tapetum lucidum, or "shining tapestry." We and other primates do not have tapeta, but many animals do, especially nocturnal or crepuscular ones (that wonderful word means active at dawn and dusk). We think that the tapetum is present to reflect back any light not detected by the retina. This gives the retina an optical second chance and almost doubles the light-catching ability of the eye, but it comes at the cost of blurring the image and scattering reflected light around the inside of the eyeball. If you need to catch brown mice on a brown background at twilight, perhaps that is a price worth paying.

So the eye is an assemblage of oddly named transparent, reflective, and sooty tissues quite unlike any other structure in the body. Having to deal with light makes the eye a very strange place. It also appears to have made the forefathers of ophthalmology seek an unusual terminology for their oft-misspelled discipline. I would love to linger over the linguistic delights of synechiae, iridocyclitis, chemosis, hypopyon, aphakic crescents, and iris bombé, but they are not strictly relevant to our story. Instead we will look at where our eyes came from, and whether or not Darwin was right to worry.

As I have already implied, vertebrate eyes, being soft and squishy things, do not fossilize very well. Because of this, most of what we know about the evolution of our eyes comes from looking at the eyes sported by today's vertebrates. In some ways they seem diverse, but to be honest they are merely variations on a basic theme. Clever and subtle variations admittedly, but not fundamental.

For example, there are variations in the retina that seem eminently sensible. Many animals, including ourselves, have a fovea or "pit" at the center of their retinas with especially tightly packed photoreceptors. These photoreceptors yield a high-resolution image of the area at which we are gazing. Some animals adapt this to their own needs—birds of prey may have two foveae, one for gazing downward and one directed at the horizon. Raptors also manage to cram in more photoreceptors than any other vertebrate—up to five times more in a patch of retina than humans. Another way to see more acutely is to make the entire retina larger, which can mean that eyes become so large that they can no longer move in their sockets, necessitating the flexible necks of owls and bush babies. Also, in many vertebrates the photoreceptors have been modified to detect color, and birds and fishes may also add colored oil droplets to them to enhance their color sensitivity. Humans are quite good at color—having three different photoreceptor types to discriminate colors—but many birds have five types, including some that also detect ultraviolet light.

The way that vertebrate eyes focus light is probably the most variable thing about them, but here I would again argue that this reflects the relatively recent acquisition of lenses and thus their relatively experimental nature. On the scale of evolutionary history, we only just got lenses, and we are still working out what to do with them. In aquatic vertebrates, the lens has to do most of the bending of incoming light to focus it on the retina, and as a result is extremely optically dense compared to the fluids around it. In contrast, in terrestrial species most of the focusing takes place at the interface between air and cornea, so the main role of the lens is just to tweak the light slightly to adjust between looking at distant and nearby objects—lenses in land species are thus not nearly as optically dense as those of fish. In all vertebrates the lens also has a secondary role of removing optical imperfections related to the differential bending of light of different colors or light arriving at different parts of the cornea. This it achieves by varying in optical density throughout its own thickness, a feat never achieved in artificial lenses. All these various demands probably explain why vertebrates differ so much in how they focus: Jawless fish and some birds have muscles to flatten the cornea; sharks and amphibians have muscles to pull the lens forward; bony fish usually pull the lens backward; reptiles have a ciliary that squashes the lens flatter; birds and mammals' ciliary stretches it flatter. And focusing can become very strange in species with particular needs. Diving birds' lenses have the greatest focusing range of all so they can see both above and below water. Some so-called "split-eyed fish" have two-part corneas with an upper part to see in the air and a lower part to see under water. Snakes' eyes are so weird and confusing that we now assume that they almost lost them during some subterranean episode in their past and subsequently had to rebuild the degenerate stumps into fully functioning eyes when they changed their minds and resurfaced. Chameleons are the only known animals, vertebrate or otherwise, to have a telephoto lens system in their eyes. Growing flatfish move one entire eye to the other side of their head so it lies alongside its fellow. In most species it slides over the top, but in some it almost defies comprehension by migrating straight through the middle of the head.

Yet through all this, the vertebrate eye remains essentially the same. We can pack the retina a little denser, or jiggle the lens about in different ways, but it seems as if the basic design of the eye cannot be bettered. This sort of perfection is no help at all when we want to know how it came to be in the first place. To do that, we need to look even further back to a time when vertebrates were not even a distinct group. If the first part of this book was about viewing the brain as geography, then perhaps this middle part is more about seeing the brain as archaeology.

So if we look at animals as a whole, what do we see? For a start, the twenty or thirty major subdivisions of the animal kingdom fall into three main groups. One third of them cannot detect light at all. Another third have special cells that can detect diffuse incoming light. The other third, including us, can form a focused image onto a sheet of light-sensitive cells. The fact that many animals can detect light without forming an image suggests that our ancestors probably detected before they focused—after all, why bother to focus before you can detect? You may already have realized that some noneye parts of your body can respond to light even though they do not form an image—the pigment cells in your skin make you tan when they are exposed to ultraviolet light, for example. Clearly one does not need eyes to respond to light.

So if light detection came before eyes, how do different animals detect light? The answer seems to be that they all do it in a remarkably similar way. In an amazingly diverse array of creatures, light is detected by a large protein molecule embedded in the surface of detector cells, with a little, light-sensitive vitamin-A-like molecule attached to it. The big protein is often called an opsin (Greek for "sight") and the little molecule is a chromophore ("color-carrier"). We humans make our own opsins, but we cannot make chromophores, which is why we need vitamin A in our diet, and why there is a grain of truth in the old wives' tale that carrots help you see in the dark. When a photon of light hits a chromophore, the molecule flexes about one of its internal chemical bonds, inducing a slight change in the shape of the larger opsin molecule that grasps it. The next steps vary among different species, but the change in the opsin protein triggers a chain reaction of chemical alterations within the cell. In vertebrate photoreceptors, for example, the flow of sodium through the cells is altered, causing electrical changes that excite the adjacent neurons connecting the photoreceptors to the brain, just as in Cajal's picture (see Figure 7.1). Thus is light detected by almost every animal that can "see."

The opsin-chromophore system appears to have been an unbeatable way of detecting light. Or at least, it is so good that once an animal is using it there does not seem to be any need to replace it. Presumably this is why it is used by everything from single-celled organisms to blue whales. Certainly it is flexible. For example, tiny changes in the structure of the opsin molecule can "tune" the chromophore to respond to different wavelengths of light. Because of this, most humans have four different opsins. Three of these are present in three discrete populations of photoreceptor cells to allow us to distinguish colors. We probably evolved this trichromatic vision to help us determine the ripeness of our fruity and leafy diet, although as I discussed in my last book there seems to be some reason why it is advantageous for many men to be deficient in color discrimination. Another way in which the opsin-chromophore system can be modified is that the chemical chain reaction that it triggers can be self-regulating—our own photoreceptors respond to bright light by becoming less sensitive and to dim light by becoming more sensitive. Because of this, we are able to see in ambient light intensities differing by a factor of as much as a billion—far better than any manmade camera. In fact the main factor limiting our dim-light vision is probably interference from the heat given off by our own bodies with the sensitive light-detection systems of the retina.

So if vertebrates have inherited the light-detecting chemicals present in most animals, then that does not really help us understand how our own eyes evolved. But what about the photoreceptor cells themselves? For a long time we thought this was something that also divided animals into two neat groups. Some animals have photoreceptor cells with little finger-like projections called microvilli, whereas others have photoreceptors with thinner structures called cilia. Vertebrate eyes fall very clearly into the ciliated camp. As we have seen, cilia are a common thread throughout our special senses. The single cilium of human rod and cone photoreceptors is a tremendously distorted thing, however, as it carries upon it multiple stacks of lamellae, or "little plates" in which lie the opsin-chromophore complexes. These lamellae are temporary structures, continually created at the base and discarded at the tip of the photoreceptor. Also, whereas the lamellae in some of the photoreceptor cells, the cones, are infoldings of the surface of the cell, the lamellae in the other photoreceptor cells, the rods, are discrete internal structures. We think this distinction may explain why rods are so much more sensitive to dim light. The lamellae of the cones have an additional task— they carry the three different opsin types that allow the detection of color.

The microvilli/cilia dichotomy seemed a very pleasing arrangement. It suggested that photoreceptor cells evolved twice in the animal kingdom and that these two events gave rise to two great families of seeing animals, even though both use the same ancestral chemical detection system. Recently, however, this idea has been overturned by the discovery of animals with photoreceptors containing both microvilli and cilia. If the two photoreceptor cell types evolved independently of each other, then there is no way that a single animal could have both. And the subsequent discovery that many microvillous receptors actually sprout a few cilia as they develop has heralded the death of the two-great-families theory. Instead, it seems that photoreceptor cells also evolved only once, and all "seeing" animals share the same basic, though often modified, design.

This unity disappears when we look at how these photoreceptor cells became arranged into eyes that form images, however. Here there are several different designs in nature and a great deal of evidence that some of them have evolved many times in different animal groups. There are two main types of eye—the compound eye and the simple eye—but within each there are sub-groups. Compound eyes, such as those of insects, consist of multiple surface lenses or ommatidia (Greek for "little eyes") that each direct light rays from one specific direction toward the photoreceptor cells. The simplest system is that each ommatidium feeds into one photoreceptor, but there are also at least three different ways in which light from different ommatidia can be bounced around and shared among many photoreceptors. In contrast, simple eyes like ours have only one hole through which light enters, but even here there is variation. First, the light entering through that hole may produce an image in a manner similar to an old lensless pinhole camera, as in the mollusc *Nautilus*. Alternatively, in a very few species of clam the light may be reflected from a convex mirror onto a sheet of photoreceptors slung within the concavity of that mirror. Finally, in order to produce the brightest, sharpest image, a lens may be suspended in the path of the light, as in vertebrates, squid, and octopi.

The eyes of humans and squid look amazingly alike at first sight. The gross structures are so similar that it is very tempting to assume that they arose from the same ancestral eye. They each have a cornea, iris, lens, and retina. Yet when we look at the microscopic structures, we make discoveries which show that they simply cannot be descended from the same eye. One fundamental difference is that the squid's retina is arranged sensibly, with the photoreceptors pointing toward the incoming light. We know this because Cajal showed us which way round nerve cells are—with dendrites pointing toward the sensory cell and axons pointing toward the brain—and Cajal worked on octopi and squid to demonstrate exactly this point. The human retina, however, is the other way round. Rather illogically, light must pass through all the nonreceptive layers before it reaches the photoreceptors (Figure 10.2a). This single difference means that vertebrate and mollusc eyes must have evolved entirely separately— we simply cannot see how an animal with a complicated lensed eye could invert its entire retina.

In fact, when we look closely at the eyes of many different animals, we find the same thing again and again—ostensibly similar eyes with niggling little differences which mean that they must have evolved independently. Remarkably, some zoologists now think that the evolution of the eye from previously eyeless ancestors occurred over a hundred times in different realms of the animal kingdom. Suddenly we have come from Darwin's worry about how eyes evolved at all to saying that they have evolved over and over again; in fact it seems extremely easy for eyes to evolve. There is even a shrimp that has bothered to evolve both compound and simple eyes. Eyes are no longer the cherished, hallowed organs of yore. Instead they are utilitarian additions that animals seem to be able to sprout at the drop of a hat.

To further reinforce the notion of eyes as easily evolved, scientists have also suggested how eyes can evolve out of nothingness. They have even developed computer simulations that show the process occurring. Most important, they have answered Darwin's fundamental worry—how each step along the way to making a finished eye can confer a little extra advantage on an animal that allows that step to be

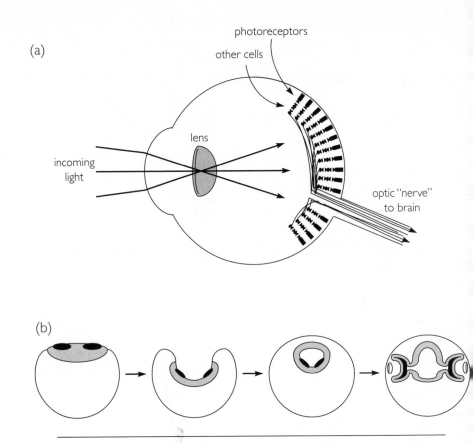

Figure 10.2. The "inside-out" arrangement of the human retina, and how it got that way. (a) Light is detected by a sheet of specialized ciliated cells called photoreceptors. Once these are activated, they induce activity in an orderly array of neurons that eventually convey the visual information back to the brain. The photoreceptors are the part of the retina farthest from the incoming light, and so all light must pass through and presumably be distorted by all the other layers of retinal cells before it reaches the actual site of light detection. (b) Four proposed stages in the evolution of our retina. Each is a cross section of the head of one of our distant ancestors. The vertebrate nervous system started off as a flat slab of tissue on the surface of our ancestors' backs, just as it does in developing human babies. Two patches of light-sensitive cells then formed on this patch, much as they do in human embryos. These eye spots formed near the surface of the nervous system slab, so that they could better catch the light. When the nervous system folded inward to create the sealed tubular nervous system we have today, the eye spots ended up on the internal surface of the brain. To get over this problem, the eye spots were pushed back out toward the surface of the head on little stalks, which eventually became the optic nerves. However, the eye spots were still deeper inside the head than the brain tissue, and so to this day the photoreceptor cells remain buried below a layer of brain.

retained by natural selection. The first animal in these arguments is, of course, blind. The development of light-sensitive cells allows it to distinguish between light and dark—the shadow of a predator perhaps, or the warmth of the life-giving sun. If these cells cluster into two patches, the organism can get a very general sense about the direction from which light comes. If those patches then become concave pits, the light cast into the different parts of those pits would allow the animal a greater ability to distinguish the origin of that light. If the pit becomes so deep that the opening is a small hole, it will now function as a pinhole camera—indeed, this is the stage at which *Nautilus* finds itself. The formation of a lens from the overlying skin can be a gradual process, because even an incomplete lens will focus light better than no lens at all. Once the lens is complete, the animal can evolve muscles to move, stretch, or squash it, and once again even incomplete structures will confer some advantage.

In a short, simple story, a blind animal evolves eyes just like us. And the plethora of eyes that are design originals in the animal kingdom suggests that it truly is that simple. Making eyes turned out to be the easiest, most natural thing in the world. In fact, it is now difficult to see why so many species with photoreceptors do not bother to evolve eyes. This is especially true now that we know that most animals with eyes use the same genetic machinery to control the formation of their different eye plans and that this machinery is lying unused in eyeless animals. Although eyes have evolved many different times, we have discovered only seven optical types of eye (if we include the telephoto chameleon). Perhaps these are the only possible ways to make an eye—at least until the eighth is discovered.

Before we leave the ironically opaque world of the vertebrate eye, there is one thing that really gets on my nerves. Why in the world is the vertebrate retina arranged so illogically? Why must light stumble through all the other cells in the retina before striking the photoreceptors? Clearly this silly system has not held back vertebrates too much, but surely it would have made more sense for the retina to form the other way around in the first place? I have to tell you, we still do not know why we have been left with this ridiculous system, but it is probably some vestige of the very earliest days of our eyes. Brain as archaeology, after all.

One theory about the illogical retina is rather pleasing. It combines much of what we know about evolution and embryonic development. The idea is that the vertebrate nervous system started off as a flat slab of tissue on the surface of our ancestors' backs, just as it does in developing human babies. Two patches of light-sensitive cells then formed on this patch, just as the progenitors of the eyes are detectable at the equivalent stage in a human embryo (Figure 10.2b). But these eye spots formed near the surface of the nervous system slab, so that they could better catch the light—and therein lies the cause of the problem. The next stage in evolution (and, as it happens, development) was that the nervous system folded inward to create the sealed tubular nervous system we have today. This was all very well, but it meant that the eye spots were now on the internal surface of the brain. To get over this problem, they were pushed back out toward the sides of the head on little stalks, which eventually became the optic nerves. However, the eye spots were still deeper inside the head than the brain tissue to which they were attached, and so to this day, no matter how complex the vertebrate eye becomes, the photoreceptor cells are still buried below a layer of brain (see Figure 10.2b).

I like stories like that. They are comfortingly like Rudyard Kipling's *Just So Stories*—an elegant tale made up long after the event. We can never test or prove them, but they appeal all the same. I have my own, different Just So story for the eye, but it is even simpler. The "Bainbridgian" theory of the origins of the illogical retina starts with the simple observation that little animals are often transparent. Transparent, that is, apart from the light-detecting pigments in their tiny eyes. I am sure that our ancient forebears were once such bloodless, pellucid beasts, through which light passed easily unless it should happen to strike their eyes. Still today, many fish fry are transparent, except for their eyes. If you are transparent, then there is really no optical reason to arrange your retina either way round—photoreceptors could face the front or the back of the eye. But there is a very good housekeeping reason to have them facing the back. Photoreceptors are incredibly demanding and wasteful things. They chew up energy like there is no tomorrow, and they manufacture and discard lamellae throughout their lives. This wasteful existence requires a constant supply of nutrients, and these were more likely garnered from body

fluids coming in to the back of the eye than from the glass-clear humors within the primordial eye itself. Also, there is a specialized layer
of cells apposed to the back of the retina whose job is to chew up and
dispose of all those lamellae cast off by the photoreceptors. These
cells are themselves pigmented, so it would be foolish for them to be
positioned in front of the photoreceptors. Better then to have the demanding photoreceptors apposed to the back of the eye, where they
can be serviced by these biological waiters and garbage men.

Now you can see why the eye has worried people. The more I find
out about the history of this worry, the more I am heartened to know
that I am not the only person who frets about this sort of thing. The
vertebrate eye is just there, with little direct evidence of how it came
to be. All we can do is make informed guesses. One thing is certain,
however. Rather than providing evidence for a benign master creator,
the eye is actually showing us that the opposite is the case. The human eye is arranged in a backward way that no intelligent designer
would ever countenance. Its positioning is the perfect evidence to
show that the imperfect, illogical bodies we have today are the result
of millions and millions of years of blind historical accident.

11

HILLOCKS, BUTTOCKS, BLINDSIGHT, AND BLACK STUFF

The Midbrain

So you now have these strange but wonderful optical devices on the front of your head. All you have to do is process the torrent of information they receive every waking second. Even within the eye itself, we can see how things are going to be: the brain is not going to cope with all that information. There is simply too much sensing going on most of the time, and most of what is being sensed is not needed. The first thing that the brain does with sensory information—visual or otherwise—is to throw away the vast majority of it. After all, how much notice do you actually take of what you see, hear, smell, taste, and feel during the sixteen-or-so waking hours of every day? We are incredibly selective creatures.

Only the visual information from the center of the retina leaves the eye unabridged. The information from all the rest is summarized by an ordered system of data compression within the retina itself. In these peripheral regions, information from more than one photoreceptor is summed into a signal in a single nerve axon—some fibers leaving the retina are carrying the cumulative output of over a hundred photoreceptors. A tremendous amount of information is thus lost almost as soon as it has been received. This means that not only is brain overload avoided, but also the optic tracts can be a sensible size. If they carried an axon for every photoreceptor in the eye, they would probably have to be about an inch thick—hardly very convenient.

It is often said that the brain somehow sharpens the edge of your field of view to make you think that you are seeing these regions more clearly, but there is little evidence for this. It is more likely that you are simply used to not worrying about the vagueness of your peripheral vision. Anyway, if you suddenly decide that you need to observe an object closely in these blurry outer reaches, you can always turn your eyes to gaze at it. One group of people who do have cause to think about this are astronomers, who spend their nocturnal hours looking at very dim objects. If you gaze directly at a star and then force yourself to look slightly to one side of it, you will find that it seems to become brighter. This is because when you gaze directly at the star, its light is focused on the cone-packed fovea, which sees best in bright light. Looking slightly to one side causes the image to fall on rods, which can detect far dimmer things. In my forays into the celestial world, I initially found it strange to force myself not to look directly at the thing I wished to see—probably because my brain thought I would not get such a sharp image—but it soon became second nature.

Once outside the eyes, a strange thing happens on the way to the brain. In all vertebrates, some or all of the axons coursing in from the retina cross from one side to the other just before they reach the brain. This takes place in the most clearly visible structure on the underside of the brain—a cruciform junction called the optic chiasm. Galen commented on this chi-shaped ("χ") structure nearly two thousand years ago, and you may recall that Cajal used it as evidence that axons can be bundled into functionally discrete cables of fibers. In nonmammals, this crossing over is complete, so the right eye plugs into the left side of the brain and the left eye plugs into the right side (Figure 11.1a). This may seem rather like the reverse wiring of the forebrain I mentioned before, but in the majority of vertebrates most of these fibers are going to the midbrain, so this may instead be an unusual feature of vision itself.

In mammals, the situation becomes more convoluted. In mammals with eyes on the sides of their heads, like rabbits, the crossing over is complete just as in nonmammals. Yet in mammals whose eyes are on

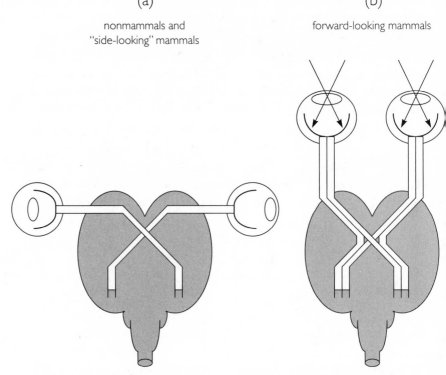

(a)
nonmammals and
"side-looking" mammals

(b)
forward-looking mammals

Figure 11.1. The optic chiasm. In all vertebrates, some or all of the axons coursing in from the retina cross from one side to the other just before they reach the brain at a cruciform junction called the optic chiasm. *(a)* In nonmammals, this crossing over is complete, so the right eye plugs into the left side of the brain and the left eye plugs into the right side. A similar complete crossing over occurs in mammals with eyes on the sides of their heads, such as rabbits. *(b)* In mammals whose eyes are on the front of their heads and point in the same direction, such as humans and cats, only half the optic-tract fibers cross over at the chiasm. The fibers that cross over are those coming from the inner half of the retina and the fibers that do not cross over are coming from the outer half of the retina. Because of this, visual information coming from the right side of the body ends up in the left side of the brain and information from the left side ends up on the right.

the front of their heads and point in the same direction, like humans and cats, only half the optic-tract fibers cross over at the chiasm (Figure 11.1b). When the chiasm is picked apart in detail, however, it becomes clear that the fibers that cross over are those coming from the inner half of the retina (the half by the nose), and the fibers that do not cross over are coming from the outer half of the retina (the half by the temple). The upshot of this selective crossing and the fact that the optics of the eye form a reversed image on the retina is that visual information coming from the right side of the body ends up in the left side of the brain and information from the left side ends up on the right. If you do not believe me, trace it through Figure 11.1b. This modification of the standard vertebrate system—the destination of visual information depending not on the eye whence it came, but from which side of the body—is probably a consequence of the way that we mammals have modified our brains. We complicated the crossing over at the chiasm by moving most of our visual processing into the forebrain, which as we have already seen is wired back to front.

After partially switching sides in this manner, your optic tract now enters the main bulk of your brain and splits up to travel toward various destinations. As I have already said, mammals are unusual in that much of the visual information now heads for the forebrain—the hindmost part of the cerebral cortex, to be precise. But in all vertebrates, including us, an important component of that visual input now makes for the upper part of the midbrain. And just as the hindbrain's history is inextricably linked with the senses of hearing and balance, the story of the midbrain is intertwined with that of vision.

The human midbrain may seem disappointingly small (see the example of mine in Figure 4.7e), but as we saw with the hindbrain, size is not everything. After the distortion of the split-open hindbrain, moving up into the midbrain is like returning to familiar territory. As pons merges into midbrain, the open roof of the fourth ventricle seals up and the brain stem temporarily narrows at the aptly cartographic-sounding rhombencephalic isthmus. Above this point the midbrain is once again a tubular structure with a central canal, just like a tubby version of the spinal cord (see Figure 9.2).

That central canal through the midbrain possesses one of my favorite brain names—the aqueduct of Sylvius. This is not just because it sounds like an edifice arching across a Roman metropolis, but because Sylvius made some rather unorthodox contributions to our studies of the brain. Sylvius lived at a time when it was fashionable to Latinize the names of great thinkers. His real name was Franz de la Boë, and he inhabited a palatial villa-cum-laboratory on the banks of the Rapenburg Canal in the pumping heart of seventeenth-century Dutch intellectual life, Leiden. Sylvius was a hugely successful and influential physician and anatomist, and he has left his name on several brain structures other than the aqueduct. We are, however, most indebted to him for his other great discovery. It had been suspected for some time that juniper berries act as a mild diuretic—they increase the production of urine. Sylvius decided that they might be useful in helping to treat kidney diseases, so he created an admixture of them with another well-known diuretic, grain alcohol. The resulting spirit was extremely popular not because it actually alleviated kidney disease but because it tasted pleasant and in sufficient quantities was able to make the drinker forget he was ill. The French word for "juniper" is *genièvre,* and so the drink was called genever. As has frequently been the case over the last few hundred years, Holland was full of English drunks, and so the drink soon came to their Bacchanalian attention, and the name was shortened to "gin." Ever since, the world has been using Dr. Sylvius' mixture to conduct what can loosely be called neurological experiments.

So Sylvius is a man whose legacy lives on in both the cocktail bar and the neurology ward—his cerebral aqueduct lying at the core of the sensory crossroads of the midbrain. Visual information flooding into the midbrain is mostly headed toward its upper part. Or is it its back part? As you can see from my MRI scan (Figure 4.7), directions become slightly confused in the midbrain because it is where humans inserted a ninety-degree bend in their brain stems when they adopted their unusual erect posture. Anyway, the upper part of the midbrain is rather logically called the tectum, or "roof." If you peer carefully at my midbrain in the scan, you will see the thin black line of my aqueduct, and to the top and right of that, my rather lumpy-looking tectum.

The size of the tectum varies enormously in vertebrates, and its size often correlates with animals' ability to see. For example, flight requires good visual skills, and so the tectum is huge in flying vertebrates—it dominates the upper surface of birds' brains. When we make plaster casts of the insides of pterodactyl skulls, we can see that the same was true for them. The size of the tectum in fish shows clearly the extent to which the midbrain is the "eye-brain"—it is enormous in highly visual species like salmon and almost nonexistent in cave fish. Really, if we look at the vertebrates as a whole, the tectum is the third great superstructure atop the brain stem, along with the cerebrum and cerebellum, and this "optic lobe" is sometimes the largest of them all.

In mammals, the tectum takes on a distinctive arrangement. Although not as large as in many other vertebrates, it appears as four rounded bumps on the top of the brain stem—upper and lower bumps on the left and right. If the cerebellum is pulled out of the way, these bumps are the most obvious features on the top of the brain stem. In fact, these bumps are the reason my tectum looks so lumpy on the MRI scan—the plane of the scan has sliced through two of the bumps. They have had a few different names over the last two millennia. They are sometimes called the corpora quadrigemina, the "fourfold bodies," or if one is stretching one's Latin, "the quadruplets." Galen was clearly less impressed by these rounded eminences and called them the gloutia, or "buttocks." The term most often used today is quite endearing—the colliculi, or "hillocks." And in one of those inexplicable convergences of nature, snakes have independently divided up their tecta into four hillocks in a way very similar to mammals.

The division of the mammalian tectum into the four hillocks is actually rather helpful to us, because each hillock has a distinct role to play. The lower two hillocks (one on the left and one on the right) are probably a modified version of the torus semicircularis, the "semicircular knoll" of nonmammals that processes information from the inner ear and lateral line system. Similarly, the mammalian lower hillocks are way stations in the chain of nerve nuclei processing and conveying sound information up the brain stem. Do not worry that we have sound information being processed in the supposedly visual midbrain—you will see that this and the next few chapters are largely a story of how the different senses converge into a complete, merged perception of the world. The upper hillocks are rather more prominent and are probably the equivalent of the large optic tectum of nonmammals. This is the place where visual information is processed, and these upper hillocks have a neatly layered internal structure suggesting that some careful ordering and filing is going on here.

Many of our less-considered second-by-second responses to light are controlled in the upper hillocks. For example, a region called the pretectum collates the input from the eyes to decide whether the irises should be dilated or constricted to allow in more or less light. It is tempting to think that we change the size of our pupils to adapt to dim and bright light, but the extent to which the pupils can be shrunk or dilated is quite limited—certainly far less than the million-or-so factor by which the light intensity decreases as we walk from a sun-drenched day into a darkened room. It is actually the retinal photoreceptors themselves that slowly adapt to changes in ambient lighting. Instead, you should think of your irises as giving you a rapid but feeble ability to move from bright to dim, or vice versa. Often we may be using our pupil size to control how much of the world is in focus at any one time, just as photographers can blur irrelevant objects in their pictures by widening the aperture. A cat dilates its pupils before it pounces—just as we do when excited—and this may allow it to blur out everything except its prey, which it can then locate more precisely.

The visual processing of your upper hillocks goes much further than this, however. Some of the layers of the hillocks construct a spatial map of your visual world, with all the things you see in their correct respective places. To do this, the hillocks must be plugged into the optic fibers in the arrangement in which they flowed from the retina. And now that there is a "picture" of the world held in these sheets of hillock neurons, this picture can be used to do some useful things. For example, the hillocks control our ability to gaze at objects and to coordinate the wandering of that gaze. This wandering is the result of sequences of rapid, jerky eye movements called saccades—a rare occurrence of a French word in this book, saccade means "twitch." If you think about it, there is some impressive data-crunching going on here, because every time the eyeballs execute a tiny saccade, the entire visual picture is shifted a little, and so the visual world must be reassessed in the fiftieth-or-so of a second before the next saccade can be made.

The spatial picture of the world held in the hillocks is also used to control some more obvious responses to the things you see. Sometimes you respond to objects that appear suddenly, or that seem to move in an interesting or threatening way, without consciously deciding to do so. If your hillocks decide that something dangerous is careering toward your head, they can make you pull away from potential danger and also close your eyes—this flinching is called the "menace response." Hillocks can also decide if something in your peripheral vision is potentially exciting and cause you unconsciously to turn your head and eyeballs toward it. To allow the tectum to control the neck in this way, there is a special bundle of fibers running from hillocks to neck rather logically named the tectospinal tract. Depending on whether the visual stimulus is nice or nasty, some of these fibers need to cross from one side to the other at the delightfully cascading "fountain decussation of Meynert."

Spatial maps are clearly what the tectum is good at. This is the site where the information from most of the nonvisual senses is rendered into a topographical "image" as well. In parallel with the visual map of the world outside, the hillocks also generate maps of the sound world and maps of the sensations of the skin. Already the senses are being superimposed, cross-referenced, compared, and merged. It hardly seems to matter what the senses are: they are all of equal value to the hillocks. Even in animals with "extra" senses— such as echolocating bats or snakes with heat-sensitive pits on their faces—it is still the tectum where these exotic senses are processed into meaningful maps of the outside world. The senses, far from becoming jumbled and confused, are each just another source of information for working out what is important. Just as we need to cower from or be attracted by visual stimuli, why should we not also respond appropriately to attractive or repellent sounds, touches, or echoes? Really, this superimposition of the senses in the hillocks is telling us what the senses are for—not to give us several discrete high-fidelity channels of disparate information, but to give us a synthesized view of the world. All the senses in the world are of little use if we cannot comprehend what they are telling us.

The upper hillocks have often been implicated in one of the most bizarre phenomena known to modern neuroscience. Blindsight will make you think again about how that gloopy mass inside your head really works. Blindsight was first recognized in people who had suffered brain damage that had rendered them blind, or so they thought. These patients had absolutely no awareness of being able to see and so considered themselves blind by any common usage of the word. Yet simple tests show that some of these people can respond to visual stimuli. If they are shown dots on their left or right side, or vertical or horizontal lines, or red or blue circles, they become understandably irritated when asked what they see. However, if they are forced to guess the side, orientation, or color of images placed in front of them, they can "guess" correctly more often than not—sometimes with a high degree of accuracy.

Blindsight is unnerving because it strikes at our very idea of what it is to be conscious. These people do not consciously "see" anything, and yet clearly they can receive, process, and respond to visual information. There is an unconscious world of vision into which they cannot force their way—there is an immovable barrier holding them back. It must be unbelievably frustrating to be able to "see" things, but not be conscious of them. We used to wonder if blindsight was simply a freakish effect of brain damage, but we are now fairly sure that it is the way our brains are meant to work. We run parallel conscious and unconscious systems for dealing with vision, but we are unaware of the latter unless the former is lost and a neurologist forces us to answer stupid questions. Yet clever tests can show that we all have blindsight—if you search the Internet, you will find online tests which apparently demonstrate that you too have it. For example, if images are flicked in front of your eyes for extremely short periods, you can interpret and react to them even though you never realized they were there—a phenomenon some claim is used in that *bête noir* of the paranoid: subliminal advertising.

As you can imagine, neuroscientists would love to understand blindsight. They have used scans to view the processes inside blind people's heads; they have chopped out bits of monkey brains to see what happens; they have even induced temporary blindness in humans with blasts of intense magnetism. Yet still they disagree about where blindsight occurs. An obvious possibility would be that unconscious visual processes take place in the hillocks of the midbrain, and conscious vision is located in the cerebral cortex of the forebrain. We typically like to think that unconscious processing occurs in the brain stem and conscious processing occurs in the cortex. After all, you do not have to think about constricting your pupils in bright sunlight, do you? But life is never as simple as that.

First of all, I slightly misled you by saying that visual information is divided into separate pathways—a conscious pathway to the cortex and a subconscious pathway to the hillocks. Although these represent the two major visual pathways, they are not absolutely separate. In fact there is a great deal of communication between the tectum and the regions of forebrain that process vision—so the boundary between the conscious and the subconscious is not at all clear. Yet experimental obliteration of visual processing in the cortex does lead to "conscious blindness," but it leaves some elements of blindsight intact. However, other studies have suggested that we have underestimated how widespread visual processing can be, and there is evidence that some regions carry out functions of which we are not consciously aware. Certainly in brain-damaged patients it has been suggested that blindsight is due to the activity not of the hillocks, but instead of "islands" of intact cerebral cortex that continue to function even though they have lost any connection they may once have had to our conscious mind (whatever that is).

Although the jury is still out on whether or not the upper hillocks are the location of most blindsight phenomena, there is one startling aspect of blindsight that does appear to be hillock-based—recognition of fear. It has long been suspected that there are brain regions which specialize in responding to emotional stimuli, especially certain facial expressions. For example, globular regions in the forebrain called amygdalae, or "almonds," become active when we see someone making a fearful expression. Remarkably, this activity is also present in brain-damaged patients with blindsight, and this has been used as evidence of a special eye-hillock-almond pathway for responding to other people's fright. Thus, it is possible to recognize the visible plight of others, even when they cannot consciously be "seen." This remarkable ability can also be demonstrated in people with normal vision. If images of frightened faces are flashed before your eyes so quickly that you cannot consciously perceive them, they can still activate your almonds. Thus, deep in your head, there seems to be an ancient, almost visceral system for responding to the terror of one's neighbor. An empathy pathway, perhaps?

Fascinating though the tectum and its colliculi may be, these hilly uplands make up only a fraction of the small mass of your midbrain. Below the aqueduct of Sylvius lies a region rather un-descriptively called the tegmentum, or "covering" (see Figure 9.2). And whereas the tectum above it is mainly involved in sensation, the tegmentum is more dedicated to motor activity—moving things. If you think back, this is a very similar arrangement to what we saw in the spinal cord: the back of the cord handles sensory input and the front of the cord deals with movement. Admittedly we have gone round a ninety-degree bend, but now the unpeeling of the hindbrain is behind us, we can see that the relative positions of the motor and sensory regions have remained the same throughout.

Good examples of the way that the tegmentum is all about movement are the neuron clusters that control the eyes. Only two pairs of cranial nerves exit the midbrain, but both have their nuclei in the tegmentum and both go to the eyes:

IV. Trochlear, "pulley": moves one eye muscle
III. Oculomotor, "eye mover": moves most eye muscles, including iris and ciliary
II. Optic (enters in front of the midbrain): visual information from the retina

The trochlear nerves are really just put there to annoy anatomy students. They are tiny. They come out of the top of the brain, unlike all the other cranial nerves. They inexplicably cross from one side to the other, again unlike all the others. They each supply just one tiny muscle that rotates the eyeball, and even that has an unnecessarily complex configuration that involves making a ninety-degree turn around a gristly pulley—trochlea is Latin for "block and tackle."

The oculomotor nerves are altogether more straightforward and important—they move most of the eyeball-moving muscles, so it is via these that the tectum causes its saccades. The left and right oculomotor nuclei are joined in the midline at a disputed nucleus—

called either the nucleus of Perlia or the nucleus of Spitzka—where we suspect that the directions of the two eyeballs are compared to calculate the distance of objects by trigonometry. As if that were not enough, the oculomotor nerves also drive the constriction of the iris and the distortion of the lens by the ciliary, and there are special nuclei set aside for this, the blissfully Teutonic Edinger-Westphal nuclei—Ludwig Edinger and Karl Friedrich Otto Westphal were late-nineteenth-century Prussian neuroanatomists.

The rest of the tegmentum is a colorful place, although any break from the buff drabness of the brain can be a sign of looming problems. On each side there is a nucleus ruber, or "red nucleus," which is an important center for coordinating motion. In the species I deal with, the ruber is king—it is probably the main controller of voluntary movement in most mammals. In humans the pyramids have usurped this role, and yet the ruber is still there, important and if anything even pinker. Even in bipedal humans the ruber seems to be there to allow us to walk with all four limbs. We use it a lot during our infantile quadrupedal crawling stage, and as adults we also use it to swing our arms when walking. All that said, we are still not entirely sure why the ruber is red—maybe it is because it has a very good blood supply.

Also in this area is the wistful-sounding locus coeruleus, the "sky-blue place." Its ethereal blue color probably results from the deposition of long chains of the chemical that its neurons release, norepinephrine. The coeruleus is in no way a restful place, however. It is probably important in driving the rest of your brain to be active when it needs to be, and it is involved in alertness, arousal, stress, and ultimately panic. Its neurons send meandering tendrils to almost all other parts of your brain to jolt you into action—for example, it is almost certainly part of the fright-recognition pathway between the hillocks and the almonds. Intriguingly, it is also important in dreaming sleep—something to which we will return briefly in the final chapter of this book. Finally, and perhaps unsurprisingly when you consider what it does, many antidepressants are thought to act on areas with which the locus coeruleus communicates. Maybe depression is when the sky-blue place darkens into twilight.

While we are in these dark, mysterious territories, it is worth mentioning another all-pervading and enigmatic structure: the "netlike" or reticular formation. It is customary for scientists to speak of the net as a vague and ancient meshwork of poorly defined cells permeating the entire brain stem. This is mainly because simple organisms often have a netlike arrangement to their nervous system, but of course this does not mean that our reticular formation must be inherently crude or primitive. Besides, there is no firm evidence to show that our net is a vestige of the simple nervous system of our ancestors. Even the name is misleading, because the mammalian reticular formation has at least thirty regions where it is condensed into recognizable nuclei—so really it is unusually complex rather than too simple. Hence we should not be surprised to hear that the net is an extremely important entity. It probably keeps us awake; it allows us to stand and then frees our limbs so we can move; it helps to control our internal organs; it controls what information reaches our conscious mind; it conveys and contextualizes pain. Not bad for a supposedly crude and archaic system. The reason I mention it here is that quite a lot of the reticular formation is clustered in the midbrain—around the aqueduct, for example, in the so-called periaqueductal gray. I will never forget that late night I was first called out to see an animal that had suffered trauma to its midbrain in a roadside traffic accident. I had been taught that midbrain damage causes somnolence because of damage to the arousal centers of the reticular formation, but the profound stupor I observed in an otherwise conscious animal was really rather disturbing. In a reversal of the usual adage, there was someone home, but the lights were not on.

Beneath the tegmentum, the midbrain flares out sideways and forms a central cleft along its underside (see Figure 9.2). In humans this flaring is unusually dramatic, and the two lobes so formed are especially prominent. From underneath they appear as two large stalks, one on either side, plugging into the hemispheres above. Because of this, they have been called either the crura cerebri or the cerebral peduncles—the "legs" or "little feet" of the brain (although "peduncles" could easily cause confusion with the cerebellar peduncles). One of my erudite colleagues—a vet, a Ph.D., and a man of the cloth to boot—rather memorably describes the cross section of the midbrain as looking like a pair of hot pants, although you may wish to check on the picture to see if you agree with him. Much of the crura is made up of the large bundles of motor axons descending from the hemispheres to form the pyramids—the main route by which the human body, at least, is moved. In many other animals the crura are smaller. Also, dangling between the crura in a rather intimate region of the midbrain hot pants is the interpeduncular nucleus of Gudden. Something of an enigma, this nucleus seems to be very ancient, and connects to another mysterious structure we will soon encounter farther forward in the brain, the habenula. To give you an idea of the strange nature of this region, when it is damaged, animals exhibit behavior called "obstinate progression," in which they are unwilling to end activities once started. For example, they may mutilate themselves by repeatedly attempting to walk through solid barriers.

Last of all in our tour of the midbrain, sandwiched between the crura and the tegmentum is a layer clearly visible to the naked eye and famous for all the wrong reasons. This is the mundane-sounding substantia nigra, or "black stuff." The very blackness of this stuff may be its most important feature. The black stuff actually has two different parts, the pars compacta and the pars reticulata—the "compact bit" and the "fishnet bit," although perhaps we should now end our dalliance with this fetishistic concatenation of little feet, hot pants, and black fishnets.

The blackness itself is conferred by neuromelanin, an aggregation of dopamine, the chemical that these cells release in the same way that the cells of the nucleus coeruleus accumulate sky-blue pigment. In a cut section of the midbrain, the neuromelanin in the substantia nigra makes it stand out as a visible stripe, although this is only true under certain circumstances. For example, when you were born there was almost no neuromelanin in your black stuff. Instead, it slowly accumulated inside the neurons until it became visible at around five years of age. As the pigment accumulates, the stripe becomes darker and darker until it reaches peak intensity in middle age. Neuromelanin is not the same as the melanin that makes your skin dark, and so it is even present in albinos. Another interesting variation is that it is not widespread in the animal kingdom—we think that only mammals make neuromelanin, and even many of them do not seem to bother. Cats and dogs make a little, for example, but it is humans that make by far the most. And there is one intriguing group of people in whom it seems to fade—sufferers of Parkinson disease.

First identified by James Parkinson in 1817, the disease that bears his name is one of the most common and distinctive of all degenerative brain diseases. It affects perhaps one in two hundred people, with over four-fifths of cases first showing symptoms over the age of fifty. Unlike multiple sclerosis, it does not seem to have a particular attraction to any ethnic, geographic, or socioeconomic group, and its pattern of occurrence can seem bewilderingly random. One straightforward feature of the disease is its symptoms, which explains why it was discovered so early and has been fairly reliably diagnosed ever since. Sufferers exhibit a gradual deterioration in the smoothness and coordination of their motor activity, so that their movements become slow and stiff, and they usually develop a characteristic tremor. They may also have problems balancing and have a shuffling walk. They may have a rather unchanging facial expression, and their speech may become mumbling and muffled. Also, they may suffer from depression.

Remarkably, we now think that all these diverse, severe abnormalities of movement result entirely from degeneration of the compact portion of the midbrain black stuff, the substantia nigra pars compacta. As I relate in Chapter 17, there are neuronal circuits in your forebrain that control voluntary movements, largely based in structures called the basal nuclei. Controlling conscious movement is no simple matter, but there seems to be a single input to this system that acts like a control knob for smoothness of movements. This is what the dopamine from the black stuff is for—modulating the slickness of motor activity taking place in the forebrain, maybe by allowing it to select decisively which groups of muscles are activated and which are left unactivated. Parkinson disease seems to occur when over 80 percent of the black stuff's dopaminergic neurons, as they are called, are destroyed. This is why these patients have such pallid strips of black stuff—it is these cells that contain neuromelanin. And we are pretty sure about the central role of these neurons in the disease, because we can mimic the symptoms in monkeys by administering toxins that we know kill dopaminergic neurons. And whether caused by natural degeneration or deliberate toxicity, the destruction of the substantia nigra control knob leads to uncoordinated, jerky activity by the motor system—neurons in the basal nuclei fire in erratic bursts; they establish abnormal oscillatory behavior; they start to fire in concert with nearby neurons when they should be damped down. So the outward disordered, shaking signs of Parkinson disease reflect paroxysmal, vibrational activity within.

Why should this system, among all others, be so very susceptible to degeneration and decay? Could this be something to do with its most distinctive anatomical feature, its pigmentation? Indeed, we are now becoming increasingly suspicious that disordered pigmentation of the substantia nigra underlies this common and devastating disease. Really, the first question we have to address is why the pigment is there at all. It is tempting to think that neuromelanin is merely an incidental accumulation of dopamine junk that could not be disposed of in any convenient way, yet many other areas of the brain cope perfectly well without obsessively hoarding pigmented chemicals. Thus, attention has turned to the idea that neuromelanin may actually have some function maintaining the well-being of the cells in which it accumulates. However, the chemistry of neuromelanin has proved difficult to study—unlike skin melanin, it makes up only a small fraction of the weight of the cells in which it resides, and within those cells it is relatively inaccessible inside tenacious fatty envelopes.

What little we know about neuromelanins tells us that while they belong to the broad group of life molecules called melanins, they seem to be a rather heterogeneous brew. Still, their membership in that group suggests that they could play some interesting roles in the brain. Melanins are chemicals at the interface between chemistry and electromagnetism, and they have even been compared to artificial semiconductors. One possible function for neuromelanin is that it may act as a nonspecific binding agent for various toxins—heavy metals, fats, pesticides, and the damaging oxygen free radicals about which one hears so much in the anti-aging cream advertisements. Either neuromelanin could be locking up these poisons in a safe, permanent toxic waste dump within the brain, or it could be absorbing transient excesses of these chemicals, only later allowing them to leach slowly and safely away. We do not know whether it is a failure of this neuromelanin protective system that actually causes the disease, or whether incoming toxins overwhelm a basically healthy system, but the end result seems to be that the black stuff neurons become vulnerable and sick. Chemicals, possibly neuromelanin itself, exude from the injured cells and may incite nearby glial cells to launch an inflammatory response that damages them further. So, once the neuromelanin system is overwhelmed, there seems to be no way back. The neurons die and the brain is consigned to a pallid, jittery, stumbling future.

Parkinson disease has found itself at the center of some angry debates in medical ethics. Treatments for the disease come in various forms, but it is those that could most honestly be described as cures that have attracted the most vitriol. At present, most sufferers are treated with drugs, and most of these act to help restore levels of dopamine in the brain. Taking dopamine tablets themselves is no use because the chemical does not reach the affected brain tissues. Instead, most patients take levodopa, which enters the brain and is converted to dopamine once inside. Unfortunately, quite high doses of levodopa are needed, and these can cause serious side-effects outside the brain, especially sickness and disordered movements: dyskinesia. To manage these side-effects, levodopa is usually administered with additional drugs that prevent it being converted to dopamine outside the brain or reduce the effects of any dopamine produced. Other, newer drugs act within the brain to enhance the conversion of levodopa to dopamine or to reduce dopamine destruction. The result is a complex and careful pharmacological juggling act, and in some patients it can be difficult to keep all the medicinal balls in the air.

In some patients a more invasive technique may be attempted, the pallidotomy. If loss of the tuning knob for the compact black stuff leads to uncoordinated activity of the basal nuclei, then why not surgically reduce the output of those nuclei? Fortunately, we know that much of the output of the basal nuclei comes via a structure we will encounter in Chapter 17, the internal portion of the globus pallidus. Thus, various heroic procedures have been invented to destroy part of that structure and alleviate Parkinson disease. Destroying a structure so deep inside the brain is a risky business, so the procedure is often only carried out on one side at a time, following an extensive imaging process to identify exactly where the globus is hiding. Sometimes pallidotomy is even conducted in conscious patients so that any adverse effects are immediately obvious. Nowadays, the old method of literally burning the globus has been replaced by a method of high-frequency electrical stimulation that achieves much the same end. Either way, the effects of pallidotomy can be remarkable, although as with many treatments of Parkinson disease they are often temporary. However, all is not lost as the globus can be destroyed a bit at a time, and some patients benefit from multiple surgeries conducted at intervals of years.

The treatments that really rile people are those that attempt to replace the damaged dopaminergic neurons. One source of cells that can do the job is the midbrain of aborted human fetuses. As we have seen, neurons from a mature nervous system cannot usually divide or regrow very well, but of course fetal brain cells spend all their time doing just that. Human fetuses have populations of proto-dopaminergic cells in their tiny substantiae nigrae just waiting to reach forward and control some basal nuclei. And indeed, when transplanted into the basal nuclei or substantia nigra of Parkinson disease patients, these cells not only survive, they send out outgrowths, make their own dopamine, and relieve the symptoms. Nothing is ever straightforward with this disease, however, and older patients often fail to respond as well as the young. Even when improvement does occur, it is usually temporary. We are not even sure where we should be injecting the fetal cells.

Of course, there is considerable opposition to using cells from terminated fetuses in this way, although I must admit that I do not share that opposition—if ever there was a good justification for using this material, then Parkinson disease is one such justification. However, fetal-cell transplants present us with another, more practical problem, because it is unlikely that there will ever be enough fetal material for all the patients who need it. A better option in the future may be to attempt to replace the dying black stuff with embryonic stem cells. Obviously a hot topic at the moment, stem cells are medically exciting for several reasons. First of all, they have the ability to specialize into any adult cell type, including dopaminergic neurons; second, once established, a stem-cell "line" will divide and grow forever under artificial conditions; and third, relatively few human embryos are needed to create an array of cell lines that can treat all of the genetically diverse human population. Thus, if we could get them to work, stem cell–derived dopaminergic neurons would have a great advantage— an infinite supply of cells from a very limited embryo resource. Steps are now being taken to use stem cells for Parkinson disease, and the signs are encouraging. Although they do not survive as well as fetal cells, they have the potential advantage of being able to seek out the areas where they are needed. One day it may even be possible to inject replacement substantia nigra cells into a vein and simply wait for them to home in on their destination.

We are currently at the stage in which embryonic stem cells have been shown to survive and partially alleviate toxin-induced Parkinsonism in primates, and so we really do think that the method could work. Of course we now have to decide whether we will accept this treatment as ethical, but to be honest, medical ethics usually seem to be simply a matter of time lag between the invention of a procedure and its acceptance by the public. Supporters of stem-cell research often use neurodegenerative diseases as the weapon of choice in their arguments, and it is easy to see why. The suffering of so many could potentially be alleviated by the use of a few human embryos— every day, many times more embryos are discarded as surplus by *in vitro* fertilization clinics. It is time to look deep within ourselves and reassess how we value the different stages of human life. Is it time to use what we know to defeat this debilitating disease from the darkest recesses of our minds?

12

STINKIN' AND THINKIN'

The Origins of the Nose

Smell is a strange thing. We know we can smell, but we do not think about it much of the time. We feel that lots of animals can do it more sensitively than we can, although we do not know whether they actually sense different things or just sense the same things better. We are very bad at describing smells—unlike seen, heard, tasted, and touched things, we simply do not seem to have the right words. Yet subtle smells can induce the strongest feelings and drives in us. Why do we react so strongly to the scent of rain falling on hot stone after a long drought, the aroma of bacon mixing with egg yolk, the scent of a curtain behind which we used to play as a child?

There is no doubt that we have found studying smell more difficult than studying all the other senses, and there may be good reasons for this. For example, we think of humans as a relatively smell-insensitive species, and in this chapter I will discuss the extent to which this is true. Certainly our ineptness at describing smells hints at a certain lack of ability to analyze smells in the same way as we analyze other sensory input, and there are good anatomical reasons for this. Whatever the cause of our difficulties, smell's resistance to analysis certainly makes it difficult to design experiments. Another challenging feature of smell is the way that our noses can detect such a huge variety of smelly things (odorants)—how can an organism with only twenty-something thousand genes to run its entire body have a system that detects millions of different smelly chemicals?

During the many centuries when all we knew about the brain was its structure, smell also seemed to be the odd one out among the senses. The nerves from the nose plug into the forebrain, and so they complete our holy trinity of senses and brain bulges—as we have already seen, vision connects to the midbrain and hearing to the hindbrain. Yet the forebrain is the most distorted and modified part of the brain in many vertebrates, and this is especially true of mammals. Even if it was originally the "smell-brain," over the course of evolution we have crammed more and more functions into this foremost part of the central nervous system—the process of cephalization I mentioned in Chapter 6. Thus, the sense of smell just happens to be flowing into the region of the brain where we have concentrated our higher sensory functions, much of our motor function, our language skills, our memory, our emotions, and many other things as well. Worse than that, some of these new, incoming functions seem to have stolen large chunks of the old smell system to use for their own ends. Because of this, much of what we like to think of as our advanced cerebral computer is really just reconditioned stink-processing equipment.

All that said, where did the sense of smell come from? Smell is often said to be our most ancient sense, but I am not sure that the evidence really stacks up in favor of that conclusion. Certainly it seems reasonable to suggest that ancient organisms learned to discern nice and noxious chemicals before they learned to detect movement and light. Yet as we will see, there is good evidence to suggest that the smell system we have now—the scientific term for smell is "olfaction," by the way—is actually more recent in origin than our visual system and possibly our hearing system as well. Similarly, the smell parts of the brain are often described as some sort of fusty, ancient backwater superseded by other, more flashy regions. And that assertion also has little to back it up. For many animals, smell is by far their most important sense, so is it likely that something so very valuable would be allowed to fossilize and become outdated?

It is probable that the vertebrate nose started life as two little patches of sensory cells on the very front end of our early aquatic ancestors, although we do not really have much direct evidence for this. Obviously most vertebrates form their nose here, but the jawless lamprey has its smell organ buried in a tunnel opening out on the top of its head, and the closest relatives of the vertebrates, the lancelets, do not appear to have specialized smell organs at all. The reason we think the nose evolved from placodes on the surface of the head is that, like your eyes and ears, this is exactly how your nose formed when you were an embryo.

The story of the development of the smell placode is rather simpler than that of the eye or ear, however. That old patch of smell-sensory cells is still there. It does not fold in on itself to make a lens or a labyrinth, but of course it is no longer exposed. It seems to be a bad idea to bear your smell placode on the front of your face, so many vertebrates have buried it inside nostrils or a muzzle (Figure 12.1). Sharks and rays often have their smell organs buried in clefts by the upper lip, one on each side. Widenings at the front and back ends of the clefts funnel water in and out past the sensory cells. Bony fish often seal off an enclosed tube through which water flows from a front intake nostril to a back outlet nostril—this is why many fish have four external nostrils. When vertebrates moved to land, it is thought that they punched a new hole from the nostril canal of their bony fish ancestors into their mouth to allow them to start breathing as well as smelling through their nostrils. The old back nostril was now redundant, although it may have been recruited to a new job as the nasolacrimal duct, which drains tears from the eyes into the nose. In mammals a further change took place when the newly punched-through openings in the mouth slid backward so that your nasal cavity now opens into your throat rather than just behind your teeth. The bony plate thus created between the nose and mouth cavities is the palate. Coincidentally, crocodiles, and to a lesser extent turtles, have independently evolved exactly the same structure.

So your smell cells are now deep inside your "muzzle." As human embryos develop, they do not play out this entire evolutionary story. Instead, slabs of tissue simply grow out from the face to enclose the nasal placodes in neatly defined nasal passages. Apart from protecting the sensory cells, this also means that we are able to control whether odorant-laden air passes over the cells at all. The aerodynamics of the

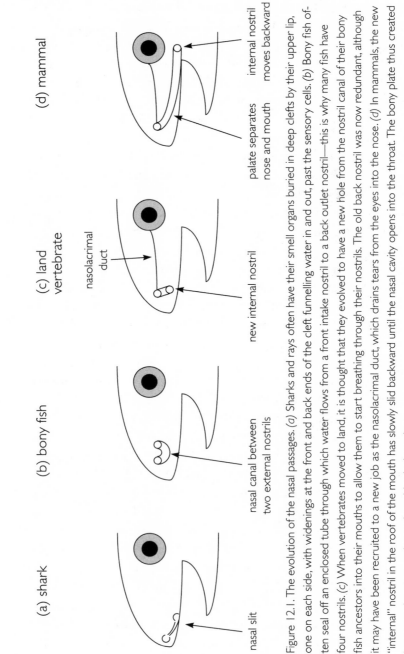

(a) shark   (b) bony fish   (c) land vertebrate   (d) mammal

nasal slit

nasal canal between two external nostrils

nasolacrimal duct

new internal nostril

palate separates nose and mouth

internal nostril moves backward

Figure 12.1. The evolution of the nasal passages. (a) Sharks and rays often have their smell organs buried in deep clefts by their upper lip, one on each side, with widenings at the front and back ends of the cleft funnelling water in and out, past the sensory cells. (b) Bony fish often seal off an enclosed tube through which water flows from a front intake nostril to a back outlet nostril—this is why many fish have four nostrils. (c) When vertebrates moved to land, it is thought that they evolved to have a new hole from the nostril canal of their bony fish ancestors into their mouths to allow them to start breathing through their nostrils. The old back nostril was now redundant, although it may have been recruited to a new job as the nasolacrimal duct, which drains tears from the eyes into the nose. (d) In mammals, the new "internal" nostril in the roof of the mouth has slowly slid backward until the nasal cavity opens into the throat. The bony plate thus created between the nose and mouth cavities is the palate.

nose are arranged so that most of the time, inhaled air simply by-passes the sensory cells and makes its way directly to the lungs. When you sniff, however, a turbulent puff of air is wafted up toward the sensory cells, which lie on the ceiling of the nasal cavity. Thus, to an extent, you can smell when you want to.

The sensory cells themselves are unusual little things. Unlike those in the eye, ear, and tongue, not only do they do the detecting, but they also themselves form the axon that sends information back to the brain. Thus, there is no intermediate nerve cell between sensory cell and brain. This may seem a rather academic point, but it actually presents the nose with unusual problems. This is because the nose sensory cells are also exceptional in that they do not last for life, but instead die after a month or two. Because of this, the sensory appara-tus of the nose must be constantly replenished by new cells. And not only must these cells be formed, but each must also send out its axon, stabbing backward into the front of the forebrain. So unlike all the other nerves entering the brain, the first cranial nerve, the olfactory nerve, is the only nerve that is continually dying and regenerating it-self. I think the idea that my nose is eternally skewering new connec-tions into the front of my brain is rather unnerving, if you will yet again excuse the pun.

This unique, continual rejuvenation of the olfactory nerve has also fired the imagination of neurosurgeons. You may remember that one of the big practical problems of the central nervous system is that when it is damaged, the nerves within it do not usually regrow. In-deed, we saw that we can track the course of cut neurons as they at-tempt to grow back toward the central nervous system, but that as soon as they reach its environs, they cease to regrow. There seems to be something frustratingly obstructive about the milieu of the brain and spinal cord. Yet we now see that smell axons are successfully growing into our brains and establishing new connections all the time, at the rate of a quarter of a million a day, in fact. It is almost as if the nasal nerves are taunting us—nerves can obviously grow into the central nervous system when they want to, but not when we want them to. Unsurprisingly, biologists have spent a great deal of time try-ing to find out what it is about the olfactory nerves that allows this constant regeneration to occur, and gradually they are discovering the reasons. Remarkably, when they extract the special glial cells that surround the smell nerve axons and inject them into a region of damaged spinal cord, they are able to initiate nerve regrowth. This is not a simple story, however, because if the glial cell sample is too pure, it usually does not work. The outlook is promising, however. Fortunately, the olfactory glial cells are moderately easily harvested (via the nostrils), and so they offer a very real opportunity for breaking the spinal-repair stalemate. Who would have thought we might heal injured spinal cords with a bit of our nose?

Back to our smelly story. How do we actually detect smells in the first place? Anatomists can see the region of the nasal cavity responsible for smell because it has a distinctive yellow-brown color. We do not know why this is, but the color is darker in species with a more sensitive nose, such as dogs. The regio olfactoria, as this yellow patch is rather grandly known, is large in dogs (up to twenty-five square inches) and cats (two square inches) but small in "microsmatic" species like ourselves, in which it is usually somewhat less than a square inch. We are no means hard done by—aquatic mammals often have extremely poorly developed senses of smell. Birds too are not particularly smell oriented, with the exception of the myopic nocturnal kiwi. Also, sometimes it is possible to be too sensitive—bloodhounds are often overwhelmed by fresh scents and may only be able to establish a directional trail after a smell has died down for a few days.

We humans manage to cram 10 million smell sensory cells into our noses, all lying under a layer of liquid in which odorants are dissolved before they can be detected. And just as light, taste, and movement-sensitive cells are derived from ancient ciliated cells, so too are our olfactory cells ciliated. It is in fact the waving cilia that actually detect the odorants in the liquid. We have now completed the set. All the major senses in the vertebrate head are based on specialized forms of ciliated cells. Why this should be, we do not know, but it has become one of the themes of our sensory story.

Scientists have argued for some time about how the cilia detect smells. An early suggestion was that the cilia bear many different types of receptor molecules, each of which binds particular odorant molecules—rather like a key fitting a lock. The problem with this lock-and-key mechanism was that the number of potential odorants seemed overwhelming—there was no way that we could make enough different receptors for all the odorants. And why can we smell artificial odorants that we would never encounter in nature? It simply did not seem to make sense. Because of these problems, other theories were put forward, most notably the idea that the nose detects the distinctive pattern of thermal vibration of odorant molecules by measuring the infrared radiation they emit. This idea that the nose "sees" odorants in infrared was an intriguing one, but it was weakened by the fact that chemicals with identical infrared properties often smell different. As a result, ever more complex ideas were suggested, including the wonderfully named "inelastic electron tunneling theory," but as time went on it seemed as if we would have to make the lock-and-key mechanism work after all.

A deft step was made in the smelly samba when we were able to look at the genes that carry the instructions to make the odorant-detecting molecules present in the cilia. Of all the genes in a mammal's body, at least 5 percent make smell receptors. I know that smell is important, but it is frankly amazing that more than one in twenty of our genes is dedicated to it. These genes all have the same basic structure, but they vary in subtle ways that allow them to bind to different sub-regions of odorant molecules. There are perhaps fifteen hundred of these genes, but we think that they evolved by repeated copying of a single ancestral gene, followed by gradual mutation of the many progeny genes until they can detect slightly different things. This has led to the largest known example of what is called a "gene superfamily."

We can now map out a family tree of the genes and even subdivide them into little "clans" of similar smell receptors. We can then look back in evolution and see how these clans of smell genes have appeared and risen to prominence over the eons. And like any genealogical investigation, it has led to considerable argument. For example, some suggest that there are clans of smell molecules which are good at detecting aquatic odorants and other clans which detect airborne smells better. Maybe the latter edged out the former as we crawled out onto the land—interestingly, some modern amphibians may have the different clans in different parts of their noses, which they can expose or hide as they move in and out of water. Another unexpected discovery is that microsmatic species do not lose smell genes, but instead simply carry them in inactivated form. For example, we humans possess roughly the same number of smell genes as mice, but maybe 60 percent are present in an inactive, degenerate form. We still lug these "pseudogenes" around, but we do not know why. One thing that does seem likely is that we have inactivated smell genes across the whole olfactory repertoire, rather than discarding entire clans of smell genes. Because of this, we think that while other animals may smell more sensitively and with more discrimination than we can, there are probably not huge gaps in our olfactory abilities.

Further investigation of the smell-gene superfamily has also shed new light on the idea that smell is ancient. When we look at the smell genes of vertebrates, they all seem to conform to the same basic plan, suggesting that they originated from the same single ancestral gene. However, nonvertebrates have smell genes that are fundamentally different. Yes, there are similarities, but those probably reflect the likelihood that different animal groups end up with similar solutions to the design problem of making smell organs. What this tells us is that, unlike vision, there is not a universal system of smell detection, but instead different animal groups had to evolve their own systems independently. Thus, our sense of smell is not an archaic relic of a time before the vertebrates had appeared as a distinct group but is instead a (relatively) recent acquisition, which appeared along with all the other things we had to develop as we became backboned animals.

Another thing we should mention before we start to follow those olfactory nerves back into the brain is the way in which the nose is always adapting to its environment. You may recall that photoreceptors in the eye have an internal system that allows them to cope in different levels of ambient light. This is why you can see in bright sunlight and in darkened rooms, yet you are still aware that it is bright or dim. Similarly, smell receptor cells adapt to ambient smelliness, but they do it so efficiently that if you are exposed to a smell for a long time, it simply ceases to register at all. I remember this most vividly when as a veterinary student I worked on a pig farm for a few weeks. As I first walked onto the farm I was almost smacked in the face by the pungent stench of the porcine inhabitants, yet within an hour or two I was entirely unaware of the smell. We simply get used to smells—our own cigarette smoke, body odor, and smelly pets. Perhaps continuous smells are of little importance to us, and we take no notice of them so they do not mask our perception of novel, interesting odors. This process of "adaptation" occurs in every human sense, but it is only in smell that it is so disconcertingly, and perhaps antisocially, complete.

The olfactory nerves do not have very far to go. Your regio olfactoria is at the top of your nasal cavity, separated from the front of the brain by a thin, perforated bony partition, the cribriform ("sieve-like") plate. The perforations are to allow the thousands of tiny olfactory nerves through—so tiny in fact that they may be the thinnest axons in the body. Once through the cribriform plate and into the cranium, these tiny fibers immediately plug into the olfactory bulbs, two small prominences on the very front of the brain. In many other species, the olfactory bulbs are much larger—so much larger in fact that they often contain their own fluid-filled extension of the ventricular system. The human bulbs are rarely fluid-filled, although they may contain at their cores thin, vestigial threads of ventricle-lining cells.

As the post-bulb destinations of smell information are varied and complex, it is lucky for us that much of the basic processing of smell goes on in the olfactory bulb. Remember that smell has always been a difficult sense for scientists to think about, and there has been much discussion about how we manage to do it. Well, it turns out that the arrangement of the bulb explains it quite well. Over the last few decades a few simple pieces of evidence have come to light. The first is that each of the 10 million smell cells in the nose bears only one of the thousand-or-so different types of smell-receptor molecule. The second is that all the cells which express a given receptor molecule send their axons to a single cluster of cells in the olfactory bulb, a glomerulus or "little ball." The bulb is therefore full of glomeruli, each wired up to just one type of receptor molecule. The third piece of evidence is that various combinations of these glomeruli then send axons to "mitral cells," which allegedly look like bishops' miters.

Now bear with me. One smell-receptor molecule can recognize the same small molecular region present on many different odorants. And conversely, one odorant molecule can have different parts of itself recognized by many different smell-receptor molecules. So the thing that defines the smell of an odorant is the combination of the different smell receptors to which it binds. Because of this, any one odorant will activate a distinctive selection of glomeruli in the olfactory bulb, and it is the mitral cells that probably recognize and interpret all these different permutations. Thus, the nose does not recognize smelly molecules in their entirety. Instead it detects little molecular regions within them, and it is the accumulated combination of all these regions—an instant molecular analysis, if you will—that allows the brain to work out what they are.

So smell is almost like an alphabetic or syllabic language. We do not have a pictograph for every word (unless we speak Chinese), but instead we construct words from letters (if we speak English) or syllables (if we write Japanese *hiragana*). And in exactly the same way, the nose does not have a receptor molecule for every odorant, but instead constructs a smell sensation from an odorant's elemental molecular parts. This is how we smell millions of different smells with only a few hundred receptor molecules—it is the combinations and permutations of these receptors that produce our sensation of a smell.

In other words, smell is completely different from the other senses and often difficult to articulate. Yet the sense of smell represents one of two crucial times that vertebrates had to develop a system to detect a potentially infinite array of alien molecules (the immune system was the other time, and if you read my previous book, *Making Babies*, you will find that it met that challenge in an entirely different way). Once vertebrates developed this amazing system of molecular codebreaking, they did not tinker with it much. And that conservatism has meant that individual glomeruli linked to individual smell receptors are located in the same place in the olfactory bulb in many different species—humans and mice, for example. It is almost as if we all have the same bulbs, grafted onto the front of our brain as an "off the rack" item.

There is one aspect of smell in which humans are clearly deficient, however. Many vertebrates—most pets and farm animals included—possess a vomeronasal organ, or Jacobson's organ, but in humans it is extremely rudimentary, if we have one at all. Identified in the early nineteenth century by Ludwig Levin Jacobson of Denmark, the vomeronasal organ consists of an additional pair of smell-sensitive organs lying on the floor of the nasal cavity beneath a wedge-shaped scroll of bone called the vomer, or "ploughshare." And just behind the upper front teeth are two tiny holes leading from the mouth up to these tiny organs. In some species the vomeronasal organs open out into the nose instead, or sometimes they open into both nose and mouth. There are various ways that odorants can be drawn into these organs—catfish pump water into them from their noses, salamanders allow water to seep to them from their mouths, and snakes smear odorants into the vomeronasal organs with their forked tongues— "tasting" the air. Many mammals perform what is called the Flehmen reaction in an attempt to suck air or saliva into the vomeronasal organ—you may have seen a randy stallion roll its top lip up to sniff a mare, or a tomcat gape amorously to scent its next romantic conquest.

The fact that many male mammals use their vomeronasal organs to detect pheromone odors released by females has led to the assumption that the vomeronasal organ is simply a pheromone receptor. Supporting this idea is the finding that the organ seems to be almost directly connected to parts of the brain involved in sexual behavior— hence the advertisements for pheromone sprays that allegedly make you irresistible to the opposite sex. However, there are many pheromone responses that are apparently mediated via the main smell organ in the nose. Also, it is entirely possible that the vomeronasal organ may pick up nonsexual smells—it is, to be honest, a difficult little thing to study.

One thing does seem clear: the vomeronasal organ is not just an irrelevant, isolated outlier of the main nose placode. It is fundamentally different and discrete in two ways, both of which suggest that it is quite an old structure—probably dating back as far as the invention of vertebrate smelling itself. First of all, the hundred-or-so smell-receptor genes for the organ are generally similar to those in the main regio olfactoria, but they are different enough to show that the organ was already separate from the rest of the nose before the enormous proliferation of the smell-gene superfamily. Also, the sensory cells here are not ciliated, but instead are crowned by thicker, fingerlike projections called microvilli. You may remember that there was also a fundamental cilia/microvilli dichotomy in light-sensitive cells in the animal kingdom as well. Here again is this same distinction, this time setting the vomeronasal organ apart from the rest of the nose.

There has been a great deal of argument about whether we have a vomeronasal organ or not. This is perhaps not surprising as it is often quite small even in animals that use it a great deal. Some human fetuses seem to have an accessory olfactory nerve, equivalent to the one that connects the vomeronasal organ to the extra little olfactory bulb which is set aside for it. Some fetuses even have the organ itself, and the same has been claimed of adults, although it lacks many of the structural features that seem to be required if it is to function. Still, it is an intriguing idea that we may all carry a tiny organ primed to detect sexy smells, despite being sublimely ignorant of its effects.

For now, I am not going to say too much more about what happens to smell information after it leaves the code-breaking computer of the olfactory bulbs. This is not because it is uninteresting or unimportant, but simply because it is disseminated all around the forebrain—a huge and complex edifice about which we have not yet said a great deal. I think it is better if, during the remainder of this book in which we tour this spectacular structure, I occasionally mention when smell information or smell-processing equipment is nearby. Smell is woven into the fabric of our fabulous forebrains, and in many cases it is that very fabric. Wherever we travel from now on, we shall never be very far away from our sense of smell.

All I will say now is that smell information goes to all sorts of different places in the brain—up to our most refined conscious thoughts and down to our most primal urges. It can tell us of the subtleties of different vintages of the same wine, or it can propel us to copulate or vomit. Smell goes everywhere. Also, the way we perceive smell seems to be uniquely plastic. How things smell and how we respond to them depends very much on our current state of mind. Smell information is constantly being altered by inputs from the forebrain, the reticular formation and the nucleus coeruleus, among others. Context is everything with smell. Does food not smell better when you are hungry than when you are cleaning a toilet? Does not the body of someone of the opposite sex smell better when it is someone you desire than when it is someone standing next to you on a sweltering, overcrowded train? There is, it seems, an appropriate time and place for every smell.

Smell pervades our brain in ways that other senses do not. The fact that we do not consciously think about it does not alter that fact. People who lose the sense of smell become depressed and may show disturbed eating behavior, even becoming malnourished or obese. Partly, this is because most of what we think of as the flavor of food is actually its smell—think how unappetizing your meals are when you have a blocked nose. Not only is smell crucial in the synesthetic creation of flavor, it also affects us in more hidden ways. When we kiss, we may be swapping sebaceous scents from around our mouths. Mice use body scents to ensure that they mate with unrelated partners, yet they prefer to nest with closely related individuals. And sure enough, women prefer the scent of men who are genetically dissimilar to them—at least they do if they are not pregnant or using the contraceptive pill. In fact, smell may be a mechanism of incest avoidance in many species, including humans. Antelope release an alarm scent from their fetlocks as they turn on their heels and run. Epileptic seizures may induce cacosmia, the hallucination of repugnant smells, but usually only if the seizures are caused by a brain tumor. Female mice may miscarry their babies if they smell a male who is genetically dissimilar to their offspring's father. Paradoxically, men may prefer the scent of a woman's vagina when she is menstruating to when she is ovulating. Wild bears may attack women who are menstruating. And women may use each other's scent to achieve the "dormitory effect"—the synchronization of menstrual cycles in women living in close proximity.

The more you know about the sense of smell, the more you realize that there is a frighteningly powerful world bubbling under our polite exterior. Evolutionary historical accident has meant that we have adapted the smell part of our brain, our forebrain, to be the crowning achievement of consciousness. But the old stinking-thinking mind is never far below.

13

INTO THE MARRIAGE CHAMBER FOR SOME SEXY SYNESTHESIA

Entering the Forebrain

Imagine that we are sailors in a tiny submarine, powering forward toward the forebrain. We are navigating through the midbrain, careening up the aqueduct of Sylvius. Above us lies the tectum, below us the tegmentum, black stuff, and the "legs of the brain." We are fighting against the current of the backward-flowing cerebrospinal fluid. Quite suddenly we reach a point where the aqueduct opens up, so that we are in a tall but narrow chamber. On our left and right we can almost reach out and touch the walls of the chamber, but the roof is high above and the floor is far below. We are in the third ventricle.

The third ventricle is the cavity inside the interbrain, the central part of the forebrain left behind when the two endbrains grew out the side (see Figure 4.5b). If we look upward, we can see the two holes of Monro, leading into the lateral ventricles of the endbrains (see Figure 4.9). We will not worry about the endbrains for now—they will get the entire final third of this book to themselves. We are currently more interested in the walls, floor, and roof of the chamber in which we are currently sailing. I have drawn a cross section of the brain through the third ventricle (Figure 13.1), but for now you should only look at the central part, the interbrain, and ignore the large lobes on either side.

The towering walls on either side of us have an upper part and

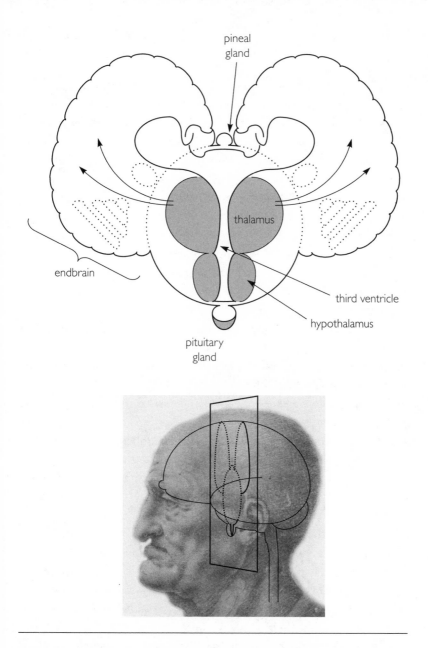

Figure 13.1. A cross section of the forebrain, with the structures of the interbrain emphasized. Below is a profile (by Leonardo da Vinci) overlain by an outline of the brain. The rectangle represents the approximate position of the section. Only the central circular region constitutes the interbrain—the larger endbrains have grown out on either side of it. At the center of the interbrain is the tall, narrow third ventricle, flanked by the two main regions, the thalamus and hypothalamus. Below the hypothalamus is the pituitary gland. At the roof of the interbrain is the pineal gland, part of the epithalamus.

a lower part. The larger, upper parts are called the thalami, which means "inner chamber," or more tantalizingly, "marriage chamber." The smaller, lower parts are called the hypothalami, or rather less tantalizingly, the "under-the-thalami." The floor of the third ventricle chamber is thin and has a fluid-filled recess in it. The roof of the chamber, above Monro's holes, has a strange mixture of structures in it and is called the epithalamus, which could mean "ode to celebrate a marriage," but I fear probably just means "over-the-thalamus." In front of us lies the boundary at the front of the neural tube, where it sealed up in the embryo. This front of the third ventricle is formed by the lamina terminalis griseum, the "grey sheet at the end," which among other things is where the mysterious cranial nerve zero, the nervus terminalis, plugs in.

So the interbrain and the tall, narrow third ventricle inside it are actually quite simple in their arrangement. One could even argue that the scheme we have followed all the way up from the spinal cord, sensory at the top, motor at the bottom, is continued here, because the thalamus is largely a sensory center and the hypothalamus could be seen as a rather special motor center. Perhaps this is stretching the scheme a little far, but I think it helps to remember which way round the different bits go. You can see the region of my interbrain on the MRI scan in Figure 4.7, and it looks like a mixed bag, largely because the structures that get cut through by the scan are as much a matter of luck as judgment.

As we enter the interbrain, the central nervous system starts to become ever more brainy and less spinal. We are no longer dealing with reflexes, running movements, and respiration, but instead we are dealing with functions like higher visual processing, sexual behavior, and knowing what time of year it is. Instead of being a place where nerves flow in and their information is immediately organized, the interbrain is at one remove from the world. It receives its information indirectly, it can select what it wants to consider, and it often has time available to do its work. We are moving from the factory floor to the executive offices.

There is one little embarrassment that I must deal with first. The whole thrust of the middle section of this book has been that each major sense connects to a brain bulge—hearing to hindbrain, vision to midbrain, smell to forebrain—and that the bulges have co-evolved with the senses they support. Also, I have just said that apart from the largely irrelevant nervus terminalis, no cranial nerves plug into the interbrain. Well, I lied on both counts. Just to make life difficult, the optic nerves from the eyes do not grow out of the midbrain; they grow out of the interbrain instead. This flagrant attempt to mess up my cozy little scheme is particularly irritating. In most vertebrates, the majority of nerve fibers leaving the eyes do indeed go to the midbrain, so why do the eyes sprout from the forebrain? I hope you do not feel that I am fudging the issue when I say that we can probably happily ignore this inconsistency. Maybe we can get around it by pointing out that the optic nerves are themselves part of the brain, and so cannot really be said to leave it at all. Anyway, eyes often have to be near the front of the head, so why not allow them to connect to the front part of the brain rather than having them trail all the way back to the midbrain? Does it really matter? No one said the brain had to follow all the rules. Let me crave your indulgence and beg you simply to forget this humiliating inconsistency.

We will start our tour of the interbrain with the roof of the third ventricle, the epithalamus. I would like to be able to say that I am doing this because this is the order in which we discussed the other parts of the brain, but that would be disingenuous. If I am honest, I would have to say that I will describe the roof first because it contains a delightfully random assortment of strange, old, mysterious structures that have long confused us. Also, despite its colorful nature, it barely gets a mention in most other books about the brain. That is the wonderful thing about writing your own book: you can concentrate on the weird minutiae that fascinate you and skim over the crucial but boring stuff.

The hindmost part of the interbrain's roof, the part just over our heads as we first chug into the third ventricle, is the subcommissural organ. Do not worry about the name—a commissure is a bundle of nerve fibers crossing from one side of the brain to the other, and the subcommissural organ just happens to lie underneath one of these crossing points. The rather vague naming of this structure does, however, reflect the fact that until recently we had absolutely no idea what it did. The subcommissural organ secretes something downward into the third ventricle, but it does not secrete a fluid. Instead it secretes a thread, called Reissner's fiber. Caught up in the flow of cerebrospinal fluid, the tip of the fiber is carried down through the aqueduct and in many vertebrates grows like a long hair until it reaches the very tip of the spinal cord.

Why on earth should the brain thread this enigmatic filament all the way through itself like this? Human fetuses have one, although it only reaches as far as the top of the spinal cord, and so it has been suggested that it plays some role in embryonic development. Yet most adult animals possess a Reissner's fiber, suggesting that it may have some longer-term role. Humans are in fact members of a select club of animals whose adults do not have the fiber, including chimpanzees, bats, and perhaps the least useful thing you will learn from this book, camels. Cattle, for example, have a big one. Theories about what the brain-hair does fall into two camps, the before-birth and after-birth theories. Before birth, the fiber has been shown to encourage normal growth of the walls of the ventricular system, perhaps by controlling cell proliferation. It is even suspected that many cases of congenital hydrocephalus may result from failed coordination of growth by the Reissner's fiber. After birth, the fiber may play rather different roles in controlling fluid flow—it may release chemicals that hold open the routes by which cerebrospinal fluid drains away into the blood. It has also been claimed to sequester chemicals like dopamine that control production of the fluid. Thus, it is both literally and figuratively a thread that passes through the ventricular system.

The structure immediately in front of the subcommissural organ has had an illustrious history. Somehow not like anything else in the brain, the pineal was once ascribed some weird and wonderful roles before we finally worked out its correct function, which is also, I am pleased to say, weird and wonderful. Some of history's greatest thinkers have agonized about this little conical knobble pointing upward from the epithalamus, often placing it at the center of their schemes of the mind. The name "pineal" comes from the Latin for "pine," as the structure was said to look like a pine nut, but Galen and the Greeks were even more specific and called it the konarium, referring to a particular species of pine tree, the European stone pine.

Before Galen, the Greeks thought that the pineal acts as a valve regulating the flow of thoughts and fluid from the front to the back of the brain—in a remarkable correspondence with what we now think its neighbor, the subcommissural organ, does. Quite reasonably, Galen did not like the valve idea because the pineal points out from the brain, not inward into the fluid-filled ventricles. Yet he was still fascinated by it, largely because it was a single, midline structure whereas, as we have seen, most structures in the brain come in left-right pairs. Perhaps because of this, early Byzantine philosophers preserved the idea that the pineal is some sort of nexus regulating the movements of thought around the brain. After all, they had just developed a fairly specific theory of localization of brain function—with imagination in the forebrain, reason in the midbrain, and memory in the hindbrain—which now needed some sort of control system to regulate flux of thoughts among these different regions.

One of the most influential thinkers in the history of the pineal was René Descartes, the early-seventeenth-century philosopher and mathematician. As a spin-off from his philosophy, he dabbled in the biology of the brain, but often with a dubious practical understanding. His theory of the pineal was developed gradually in two of his greatest works, *The Treatise of Man* and *The Passions of the Soul*. Descartes had decided that people resolved into two parts: a body, which is merely an unfeeling machine; and a soul, which must control that machine. It was Cartesian philosophy that allowed scientists to carry out horrendous experiments on conscious animals for centuries, because, so the theory went, if animals do not have a soul, then they can no more feel pain than can a machine. Descartes needed a route by which the soul could harness and control the body, and perhaps drawing on his Byzantine forerunners, he chose the pineal. His choice may have resulted from a misreading of contemporary anatomy—he claimed that the pineal is where the pathways from the sense organs flow together to create motivation and thus effect actions. So the pineal became, if not the seat of the soul, at least its intermediary: the orchestrator of the brain that physically tilts toward regions of the brain as they are needed, just as a conductor imperiously jabs his baton at the woodwind and brass.

Later anatomical studies have shown that Descartes was more wrong than he could possibly have imagined. Not only is it false to say that most sensory and motor nerve fibers experience a confluence in the pineal, it turns out that the pineal has no direct connection with nerves from any part of the brain, in mammals at least. It is that self-contradictory thing: a part of the brain with no nerves at all. It is remarkably disconnected. Its only direct link to the rest of the brain is that it just happens to be glued on the top of it. Yet when zoologists started to look at the pineals of other vertebrates, they found something remarkable—in these animals the pineal seems to be a visual sense organ. Detailed study of the pineals of amphibians and fish—especially jawless fish—showed that the pineal contains photoreceptors with stacks of light-sensitive lamellae, just as we saw in the rods and cones of the eye. Do these animals really have a third eye on top of the head? What do they use this amazing thing for? And why do we no longer have one?

Early studies of the function of the pineal were confusing. It had been known for some time that pineal tumors lead to abnormal development of the human reproductive system, especially precocious puberty in boys, yet when pineal extracts were administered to frogs, they caused a transient blanching of the skin. What could those two effects have to do with each other? An important breakthrough came when meticulous studies of the connections of the human pineal showed that it is connected up to photoreceptors, but not its own. In humans, a tiny fraction of the visual information reaching our eyes is conveyed to the pineal by a spectacularly circuitous route. First, it passes through a chain of neurons in the lower part of the interbrain and then down the brain stem and the spinal cord of the neck. It then emerges from the spinal cord around the level of the upper chest and passes through a few ganglia on its reascent up the neck. Once back in the head, the nerve fibers track along the blood vessels that supply the pineal, and once there they terminate in close association with its cells.

I can understand that the human pineal needs to get its visual information vicariously from the eye because the mammalian skull became impermeable to light, but why the pathway from eye to pineal has to be so tortuous I do not know. Surely, it would have been much easier simply to send a few nerves up to the pineal from the optic tract? After all, the pineal does have nerves going into it in human fetuses, but these regress around the time of birth. Also, the visual nature of the pineal may have been unexpected, but it is bewildering that it should take on religious connotations. The tenets of theosophy, a religion that started in the nineteenth century, state that the pineal is a degenerate relic of our third, spiritual eye, related to the third eye of the Hindu god Shiva. The pineal was thus elevated to the status of the organ of spirituality in humans, despite its diminution from its proud state in our presumably more perfect forebears. As further evidence of this descent from a perfect spiritual past, theosophists also point out, correctly, that the pineal accumulates mineralized particles throughout life, supposedly paralleling its spiritual fossilization. These mineral deposits are sometimes mysteriously called "brain sand," but I prefer the term acervulus cerebri, which descriptively refers to the dots of mold that grow on food as it ages—moldy pine nuts, one must assume. Yet it must be admitted that modern science has no better explanation for the mineralization of the pituitary, although we do find it useful when it renders the pineal visible on X-rays.

Only relatively recently have we worked out what the mammalian pineal is for. It turns out to be an important timekeeper in many species, although elephants, sea cows, and sloths apparently function perfectly well without one. The key to the pineal came when it was discovered that it is a gland that secretes the chemically unusual hormone melatonin, named after its ability to control skin pigmentation in amphibians. The pineal secretes melatonin at night, when the world is dark, and switches off during the day, so it is thought that the hormone controls many of the changes our bodies undergo during the twenty-four-hour cycle of the day. It may, for example, be responsible for the fact that we like to be active during the day and sleep at night. (It also causes twenty-four-hour variations in the rate of secretion of cerebrospinal fluid, and is in fact one of those chemicals that Reissner's fiber may soak up). We think that melatonin runs the clock that must be reset when we fly a long way east or west, resulting in jetlag. Some people take melatonin tablets when they fly.

Keeping track of the time of day may not seem too great a feat, but melatonin also allows us to do something considerably more impressive—it tells us the time of year. Since time immemorial, we have wondered how animals living in temperate or arctic environments seem to know exactly when is the best time of year to breed, to change coats, to cast their antlers, or to stock up for the hard times ahead. Some species can consistently pitch these seasonal decisions with an accuracy of just a day or two. And now that we know about melatonin we can see that it is the pineal, that indirect third eye on the sky, that allows them to achieve such amazing precision. Because the pineal releases melatonin at night, levels of the hormone are highest in winter, when the nights are long. They decrease in spring as the nights shorten, are lowest in summer, and increase once more in the autumn. And many parts of a deer, a sheep, or a stoat's body— often the brain itself, in fact—can detect this waxing and waning of melatonin levels with the seasons. So the pineal is not our soul; it is our calendar.

Melatonin and the pineal probably also explain a condition called Seasonal Affective Disorder, a depressive condition with the inappropriately jokey acronym SAD. I get SAD, and I expect many of you do, too. Every November, I feel a bit bleak, a bit dull. Once I had experienced SAD for a few years in a row I suddenly realized what was causing it, and that realization makes it much easier to deal with. We are not entirely sure what causes SAD, but it is probably a normal change we are destined to undergo in winter. Maybe we are meant to be a bit drab as the nights draw in, and you may want to speculate about why that is. Anyway, a phenomenon caused by lack of light seems to be best treated with light, and many sufferers benefit from sitting by artificial lights that mimic daylight. People with SAD as mild as mine often feel better once they understand what is going on. Maybe I just have a little too much mold on my pine nut.

I am fond of the pineal, but we must continue. The third structure in our forward clamber over the roof of the interbrain—in front of the subcommissural organ and the pineal—is the parapineal, which simply means "next to the pineal." And bizarrely here we have yet another eye, a fourth one to add to the pineal and the big beautiful pair on the front of your head. I have to admit that the human parapineal does not actually add up to much and certainly does not qualify as an eye. However, in many familiar vertebrates the parapineal often forms a more convincing eye than the pineal. It is often well developed as a "parietal eye" in frogs and lizards, although it has disappeared in snakes, presumably during the temporary sojourn underground that we saw also cost them much of their "other eyes." In one reptile, the rare and unusual tuatara or *Sphenodon* ("wedge-tooth") of New Zealand, there is even a neat little socket in the top of the skull through which peers a little eye, complete with cornea, lens, retina, and "optic" nerve, though covered by opaque scales as the tuatara matures. What is up above that is so important to tuataras, I could not say.

In vertebrates as a whole, it seems that it is the parapineal that often makes the upward-pointing eye, and that it is the pineal that converts this visual information into a hormonal signal, although it can sometimes be difficult to tell what organ does what. There is one creature, the jawless parasitic lamprey, that has both a pineal and a parapineal eye. It has been suggested that the two may have originated as a symmetrical pair. Yet in most vertebrates, whereas the parapineal usually grows out from the left side of the epithalamus, the pineal is resolutely symmetrical and loyal to the midline. But therein lies an interesting possibility. The parapineal seems to be an inherently asymmetrical structure, and even when poorly developed it connects mainly to the left side of the brain. We now think that the genetic processes which induce the wonkiness of the parapineal are similar to those which place your heart on your left side. And it now seems likely that the left-sidedness of the parapineal is the trigger that induces asymmetry in many brain structures. As we will see in a later chapter, there is considerable left-right inequality in the human brain, and it is a challenging thought that the otherwise trivial parapineal is the seed that precipitates these intriguing asymmetries.

In front of the parapineal lies a fourth structure in this queer assortment, the habenula. The name simply means "holder," because this region looked like the stalk of the pineal to those ancient anatomists who thought the pineal so important. Yet the habenula has no neuronal connection to the pineal and instead probably has more to do with our responses to pleasurable stimuli. It is a collection of little nuclei, often asymmetrical due to the influence of the parapineal, which originally may have been a major route by which information passed from the front of the brain to the back—strangely similar to what the Byzantines thought the pineal did. Much of the information was smell-related. A large input to the habenula still arrives from the olfactory bulb by way of the stria medullaris, a streak of nerve axons running up the walls of the third ventricle like some diagonal railing. In many species, most likely including humans, the smell information is sent on from the habenula to the tectum of the midbrain, where it is compared to all the other senses being superimposed and compared there.

You can probably imagine how a system connecting the nose to regions that generate responses to sensations might be slowly adapted to roles in learning, memory, and some fairly primal behaviors—mothering, sex, and suckling. After all, we are evolved from simple little creatures who learned what they liked and spent their time sniffing out more of it. Probably because of this, the habenula is now one of the most important parts of the brain for controlling our responses to rewards. Much of this brain reward system occurs as an interaction between the habenula and the interpeduncular nucleus of Gudden that you may recall we found located in a rather private region of the midbrain hot pants. These two regions are connected by a bundle of fibers either called the rather ugly habenulopeduncular tract or the more sophisticated-sounding fasciculus retroflexus. This axon pathway is not only present in almost all vertebrates, but its importance is also indicated by the fact that it is one of the first tracts to form in human embryos. It has not entirely lost its old, simple role of pursuing smelly food, however, as the brain reward system is probably still the master controller of suckling.

As we evolved into more sophisticated beings, the rewards we sought became more complex and even abstract, yet still the brain reward system controls our lusts and desires. Instead of hunger and satiety, we have depression and contentment, aimlessness and fulfillment. One remarkable recent finding has been the severe and extremely selective degeneration of the fasciculus retroflexus that occurs during addiction. Although this has been most studied in nicotine addiction, damage to this bundle of fibers is also implicated in our responses to a surprisingly wide range of substances, including cocaine and ecstasy. All of these drugs seem to tap in to our perception of pleasure and our wish to repeat pleasurable experiences, and many of them are extremely addictive as well. Are we tickling our fasciculus retroflexus when we take these drugs, or is it their destruction of the pathways that normally moderate our responses to pleasurable things that causes addiction? Is this why for some people these drugs can become the sole reward they seek—their main reason for living? Is the drug somehow even assuming the compulsive lure of mother's milk?

Structures five and six lie still farther forward in the epithalamus, but we will not say much about them here. They are rudimentary if present at all in humans, and in the creatures that have them they are poorly understood. The fifth is the dorsal sac, a dead-end pouch off the third ventricle, which may have a glandular function. And the sixth is the paraphysis, an ingrowth of the roof into the ventricle present only until infancy in humans. In other species the paraphysis may be an energy store, or it may control calcium metabolism, but in humans it only comes to prominence when it becomes cystic and blocks the flow of cerebrospinal fluid. So there you have it—the epithalamus. And what a strange assortment it is. Cruising along the top of the interbrain we have found something that spins a thread which extends down the brain, a hormonal clock and calendar, an eye on the top of the head, the seat of addiction (maybe), and a couple of ill-defined glandular things. I think that the epithalamus is the very best evidence we have that our bodies were formed by millions of years of blind evolutionary stumbling, as any sentient designer who created such a quirky mish-mash of junk would hardly qualify as a supreme being.

Beneath the menagerie of the epithalamus, things become more ordered and mainstream. Below the roof of the interbrain, the upper sections of the walls are made up by the thalami. Neuroscientists often speak of the thalamus, but in reality there are two of them, almost entirely separated by the tall, narrow chamber of the third ventricle. Together the thalami make up the bulk of the interbrain and form an important component of the sensory system, thus continuing the theme we first encountered in the spinal cord—sensory on one side and motor on the other. Yet here I feel rather torn. The thalami are large and important, but I simply cannot get very excited about them. Just like the couple you feel you should invite to dinner, the thalami are worthy but, in my opinion, dull. It is at this point that I will exercise the prerogative of the writer of popular science and say that I intend to treat the thalami lightly. They are a crucial part of your brain and have a delightful etymology, but I will satisfy myself with a brief assertion of their function and a couple of their peculiarities. Even their internal geography, with the notable exception of the Field of Forel, is boringly named. Little use to us, then.

The main function of the thalami is often said to be that of a relay station for sensory information traveling up to the cerebral hemispheres. All those sensations that we have seen rattling up through the brain stem make a final pit stop here on their way up to the hemispheres and thus, presumably, to our conscious awareness. Yet this is an oversimplification. First of all, the thalami also play a role in movement, alertness, and other nonsensory functions. Second, it seems that the thalami are not simple sensory junction boxes at all and that a good deal of sensory processing and interpretation goes on here. Much of what we know is based on studies of the thalamic regions involved in vision, the lateral geniculate ("knee-bend-shaped") nuclei, where we know that visual information is being teased apart into different components. This teasing apart of sensation is something to which we shall return, because it seems to form the basis of our understanding of the world. For example, what emerges from the lateral geniculate nuclei are two discrete parallel outputs, one carrying information on color differences, for example, while the other carries information on changes in the visual image over time, among other things. The thalamus is starting to dissect our sensations into separate streams before they reach the cerebral hemispheres.

Having said that sensory information coming up from the brain stem is relayed and analyzed in the thalami, it must be mentioned that not all this information comes in from the brain stem. As I have noted, smell is unusual because it enters the forebrain directly—the forebrain is sometimes thought of as the "smell-brain" after all. Because of this, a large proportion of incoming smell information can enter the cortex without passing through the thalami. This lack of thalamic integration has been claimed to underlie some of smell's more unnerving features. Does it somehow access our conscious mind in a kind of raw, unadulterated state? Do smells induce such strong feelings and emotions because they have not been tempered by the cold light of thalamic analysis? Although this may be overstating the case, it is clear that smell does have preferential access to the cortex and that this access may explain some of its unusual features. For example, smell is especially potent at inducing *déjà vu*, the eerie feeling that one has experienced something before without being able to remember what it is. Does smell information reach some deep, emotional core of memory before we have had a chance to search and cross-reference it in our more analytical memory banks?

Another intriguing feature of the thalami is that they differ between the sexes, but in a strangely crude way. In most people, the two thalami bulge slightly into the space of the third ventricle, and in some individuals they bulge so much that they actually touch each other. This contact of the two thalami in the midline is called either the massa intermedia or the interthalamic adhesion. It has no functional significance as no nerve fibers actually cross from one side to the other. It is simply the result of the bulginess of the two thalami. Just because two overweight people can get stuck trying to squeeze out of the same doorway does not mean that they are communing in some special way. Yet for some reason, the presence of a massa intermedia is more common in women. We have been looking at the central nervous system for some time now, and this is the first difference we have seen between women and men. As differences go it is a rather disappointing one. Women are more likely to have chubby thalami that brush against each other but make no functional connection. This hardly seems a very specific sex difference, and we still cannot explain it. But as we will soon see, it is a tentative forerunner of more striking differences between the sexes in the next and final part of the interbrain.

That next part lies below the thalamus and thus is unimaginatively called the hypothalamus, the lower half of the walls of the third ventricle. And just as I simplistically claimed that the thalamus is a sensory area, I am probably stretching the point even further when I claim that the hypothalamus is a motor region. Although it is very much a part of the brain involved in doing stuff, it is not responsible for moving large, obvious things like arms and legs, but is instead more subtle in its control of the body. Rather than moving muscles and bones, the hypothalamus is the master controller of most of our more visceral functions. It is, all in all, a rather primal thing: if we are cold, it tries to warm us; if we are hungry, it tries to make us eat; if it is time to fight, then it drives our aggression; and if the time is right for making sweet love, why then it tries to make us do that, too.

This view of the hypothalamus as controlling our basic "vegetative" functions is mirrored by the theory that among all the parts of the brain, the hypothalamus may represent the oldest part of all. If a brain is there to allow animals to make sensible responses to its environment, then this region does exactly that. All vertebrates have a hypothalamus, and our close relatives, the lancelets, also have something that seems to correspond to it as well. Even some invertebrate wormy things have a cluster of nerve cells on the underside of the front of the nervous system that seems to translate environmental cues into sensible biological responses. A kind of proto-hypothalamus, perhaps.

A great deal of information flows into your hypothalamus from just about every part of your brain and body. After all, if your hypothalamus is to decide what you need to do, then it must have a good general appreciation of how you are. Once it has decided what to do, then it can guide you by two different means. First, it can send out signals to other parts of the nervous system to control aspects of your behavior—for example, if you are cold, it encourages you to escape the wind and lie in the sun, or constricts the blood vessels in your skin to reduce heat loss. The other way it can help you is by releasing hormones—once again, if you are cold, these make you seek and digest energy-rich foods, or ensure that energy-laden molecules such as glucose are pumped into your bloodstream. I realize that these responses may seem simple, but they are extremely strong drives, and without them you would soon die.

To exert its hormonal control, the hypothalamus has had to evolve a little accessory, the pituitary gland. Brains do not seem to be very good at directly secreting things into the circulation, so they grow little extensions to help them do it—the pineal was one example. All vertebrates have a pituitary gland, and ever since Galen we have suspected that it was responsible for secreting something into us. The name comes from the Latin for "phlegm" because the gland was claimed to exude this one of the four vital humors into our system. The pituitary is also less inspiringly called the hypophysis, or "below-body," because it dangles underneath the hypothalamus on a stalk called the tuber cinereum, "the ashen hump."

Reflecting the fact that the pituitary is a major route by which our brain communicates with our internal organs, so is its embryonic development a story of interaction between brain and offal. As you developed in the womb, the hypophysis grew down from your brain. As it was doing this, directly below it appeared an upward outpouching of your throat, named Rathke's pouch after its nineteenth-century German discoverer, Martin Heinrich Rathke. The hypophysis grew down and the pouch grew up until the two fused in a unique union of mind and body (except in those idiosyncratic little lampreys, where it inexplicably grows in from the top of the head, along with the nose). The two components of the pituitary are distinguishable throughout your life. The brainy part releases one hormone involved in controlling your water metabolism and a second hormone that causes birth contractions, milk secretion, and orgasm. And under direct control of the arcuate nucleus of your hypothalamus, the throaty part secretes several hormones that are key controllers of your thyroid gland, your adrenal glands, your metabolism and growth, and your sexual organs.

There are many nuclei in the hypothalamus, but the preoptic nuclei are especially interesting, and they have certainly been the most newsworthy. Their name is arbitrary, simply referring to the fact that they lie just in front of the optic chiasm as it crosses over under this part of the brain. Yet their job is very exciting—they seem to control sexual behavior. And although it has been known for some time that this region is important in coordinating distinctively male or female behavior during copulation in lab animals, it was only recently that scientists began to wonder if this is the region of our brain where our sex is.

As I described in my last book, there is a single genetic switch that decided whether or not you made testicles or ovaries as you developed. After that, it was the presence or absence of hormones produced by testicles that drove you to be a particular sex—if you had them, you became a boy, and if not, a girl. To cut a long story short, those hormones decide whether you have male or female genitalia and internal plumbing and whether you develop a typical male or female body shape at puberty. Yet people often forget that the most important organ which must be told what sex you are is the brain. Your brain will be responsible for controlling your sexuality, so at some point it must have your sex programmed into it. And from this idea has sprung the concept of "brain sex"—the sex that you "think" you are. And as any transsexual will attest, this does not necessarily have to correspond to the sex of the rest of your body.

Because of the central role of the brain in sexuality, scientists have spent a great deal of time looking for differences between the brains of women and men. Apart from differences in total brain size, which we will think about in a later chapter, we have already seen that more women than men have a massa intermedia—hardly a striking and definitive difference. Yet meticulous anatomical studies of rodent brains have showed that there is one area of the brain that exhibits striking differences between the sexes, to an extent that suggests that it may be the very region where our brain sex is defined. This region has become tagged with the name "sexually dimorphic nucleus" of the preoptic area. It is approximately five times larger in male rats than females. And it has even proved possible to demonstrate that the size of this region can be altered by varying the amounts of hormones in the blood around the time of birth. These studies have told us that there is a short time window during which male hormones can act on the brain to change its very structure and decide an animal's sexual behavior in years to come.

As usual, however, things became less clear when these studies were extended to humans. The differences between the sexually di-morphic nuclei of men and women are less dramatic than those of male and female rats. Also, it is difficult to carry out definite experi-ments in human sexuality. Unlike rats, in which there are certain be-haviors that are specific to males and others that are specific to fe-males, human sexual behavior is much less clear-cut. Furthermore, the extent to which sexual behavior is controlled by our hormones and the extent to which it is "hard-wired" in our brains probably dif-fers from our rodent cousins. People's sexual response is strongly in-fluenced by their view of their own bodies, which makes it difficult to differentiate clearly between the sex assigned to the brain and the sex of the rest of the body. Yet despite decades of argument, it is now widely accepted that differences in the size and structure of tiny clusters of cells in the preoptic area truly are in large part responsible for our sexuality.

To add to the controversy, some researchers claim to have detected anatomical differences between the brains of heterosexual and homosexual men. The area at the center of this argument is the previously obscure third interstitial nucleus of the anterior hypothalamus, which in homosexual men is alleged to be more similar in size to that of women than heterosexual men. These studies have led to some predictably vitriolic comments, however. Some have challenged the scientific validity of these studies, such as the comparability of the groups of people involved. Others have argued that it is not possible to draw simple conclusions about the anatomy of something as complex as someone's sexuality. All in all, the whole issue seems to make many people charmingly defensive. However, why should there not be a great deal to be gained from these studies? Would it not be worth knowing that homosexuality was hard-wired into the male brain before or around the time of birth? Then we could happily ignore the conservative voices who tell us we must shield our sons from homosexual influences as they grow up (science has, as yet, little to tell us about female homosexuality). Also, from the scientific point of view, I would be fascinated to know why such a large segment of the human population does not wish to take part in sexual activities that produce children—something that runs completely counter to what Darwin would have us expect. In fact, one of the most fascinating aspects of homosexuality is that it is not some crude failure of sex-assignment. The vast majority of homosexual men are not men who feel that they should be women—they are men who want to be homosexual. Where in the tangle of preoptic nuclei could that conundrum lie?

The idea of brain sex has other implications for us, no less important. If behavior can be built into our brain structure according to our sex, then what else can? A second essential difference between the sexes is hard-wired into another region of the hypothalamus by birth, and it is to do with secretion of hormones from the pituitary. In men and women, the gonads are controlled by the same pair of pituitary hormones, and yet the way in which the hypothalamus controls those hormones is completely different. In women, there comes a time every month when the hypothalamus must respond to estrogens from the ovary with a dramatic, uncontrolled escalation of hormone secretion. This runaway secretion of hormones is what drives the process of ovulation so essential to female fertility. In men, however, there is no need for this sudden, spasmodic production of hormones—men do not ovulate or cycle—and so their hypothalami are entirely incapable of the runaway effect.

For centuries, scientists have studied differences between male and female brains. Initially this was largely focused on attempting to show that women are intellectually inferior to men, thus justifying their lower status in society. More recently attention has turned to studying relative differences in ability at different skills. For example, it is often reported that women are better at verbal skills and men are better at visuo-spatial tasks. Some people resent these generalizations, and it must be said that the evidence for some of them is flimsy. These contested differences aside, there are some differences between the sexes that may be germane to our current assault on the senses. These differences concern smell and pain.

In many species, the greatest between-sex differences in brain anatomy occur in the smell system, and this is often explained as reflecting the need of the members of one sex to find, or choose between, the members of the other sex. It has proved difficult to find similar differences in humans, but then again we do have a rather feeble smell system. There are frequent anecdotal claims about the differences between how men and women perceive smells, but it is difficult to study this rather subjective area. In contrast, there are several well-conducted studies showing that men and women perceive pain differently. Some have suggested that women are more sensitive to long-term pain than men, although mothers reading this book may wonder how this tallies with the fact that it is women who undergo childbirth. The discrepancies may be caused by differences in circulating sex hormones, and it is for that reason that women's sensitivity to pain changes throughout the menstrual cycle. Most remarkably of all, modern imaging studies of pain-elicited brain activity have shown that men's and women's brains often respond to pain to different extents or even in different regions.

What all this shows is that sensation is an inherently subjective thing. All the parallel pathways of sensory information ascending up to the brain can be modulated or interpreted in different ways, depending on what sex we are, what our hormones are doing, and presumably innumerable other factors as well. What we perceive could depend just as much on who we are and how we are feeling than on what we have actually sensed. And the tectum of the mid-brain showed us that life is even more complex than that, as senses are superimposed and compared and merged in ways that allow us to wring every last snippet of salient information from them. Are the senses really ascending to consciousness in neatly segregated streams, or are they overlapping and jumbled in a way that depends on our mental state and the way in which the brain happens to be wired? To bring our ascent of the brain stem to its close, to complete our grand tour of the senses, I would like to finish the middle, sense-oriented part of this book by looking at a wonder of nature that may show us where all this sensory information is going.

14

WHY IS "D" BROWN?

When the Senses Mix

Some people will see this book differently from others. When I read it, I see a series of little black glyphs that I quickly translate into letters, words, and then concepts. Most of you will do the same. Yet a few will see each letter or word glowing with its own color—either the characters themselves will appear colored on the white page, or a colored mist may overlie or underlie each word. The briefest glance at the page will reveal a vibrant tapestry of color, adding richness and depth to the act of reading. To people with synesthesia, the world in which the rest of us live would seem incredibly drab.

Many of us have been struck by a particular question, often in the few years of childhood between self-awareness and puberty. Just for a while we wonder if we see the world the way everyone else does. When we see a scene or smell a meal, do we all perceive the same thing, or do we all have a different take on reality? When I see green grass, do I see the same green as you? As children, we are naïvely reassured by the fact that everyone describes these sensations in the same way, not yet realizing that we may simply be using language as a common currency to create consistency between the differing ways that different people sense the world. But after a while we lose our inquisitiveness and accept that everyone probably sees, hears, and smells the same things in the same way. The ancient Greek philosophers worried about just this question and it even has an important sounding name: sensory relativism. But science has shown that some of us do not see the world the same way as everyone else. Synesthetes do not lack something that the rest of us have, in the way that color-blind people do. Instead they have something extra. And they usually grow up without realizing their unusual gift.

When a synesthete senses something, the sensation evokes a second, different sensation that to the rest of us seems nonsensical. In the commonest form of synesthesia, individual letters or words are perceived as being colored with distinctive hues. This superimposition with a second sensation is extremely rapid and very consistent. A synesthete for whom writing evokes colors sees the colors of the words on a page even before they are consciously aware of having read those words, and they rarely get it wrong. If asked to search for a word in densely packed text, they may well find it quicker to hunt for the color than the word. These people cannot switch off their synesthesia—they cannot read text without seeing the colors. They often describe their gift as helpful, and when they are asked to think about it, they say that they would rather retain it than lose it. And many of them can describe the moment when they suddenly realized that they were unusual in their abilities.

Writing that evokes colors may be the commonest form of synesthesia, but there are many other possibilities, some of which bridge different senses. A similar form is speaking that evokes colors, in which flickers of light appear when people hear certain words or names—some even see a colored ticker tape of words pass in front of their eyes. Written or spoken words can also induce tastes, smells, textures, and shapes in the synesthete. Some people see colored auras or haloes around the heads of people they recognize, a possible explanation for some religious experiences. Very rare individuals, when they see someone else being touched, experience a similar tactile feeling on their own skin. Once again, the evocation of synesthetic experience is rapid, involuntary, and consistent.

In all these diverse forms of the phenomenon, there are strands that occur again and again. First of all, synesthesia only works in one direction. For example, if a synesthete for whom words evoke colors is presented with colors, they do not "back-evoke" words. Second, some people see their evoked sensation as taking place in the outside world, such as the colored ticker-tape example, whereas others think of their synesthetic sensations as taking place inside the head. Third, synesthetes do not get confused by their gift. If presented with colored text, they can reliably distinguish the true printed colors and the colors evoked by the words. Finally, whereas the evoked sensation can be specific and complex—the taste of a particular meal or a shimmering, triangular blue shape moving upward and to the left, for example—the evoking sensations are often very simple: words, letters, or numbers. Tellingly, the strongest evokers of synesthesia are often words or characters that form part of an ordered sequence: numbers, letters, days of the week, months of the year, names of family members.

Related to this last observation is a phenomenon called number forms. Some individuals project the integers 1, 2, 3, etc. into the space in front of their bodies. For example, they may see 1 to 10 running in a row from left to right by their left arm, 11 to 100 in a rectangular array in front of them, and 101 to 200 in a square on their right. Many of these people are able to use this mental construct to carry out arithmetic tasks and often find it easier than using the methods available to the rest of us. The reason that I have mentioned number forms is that their creators are often also synesthetes in other ways, although that is not necessarily the case. The number lines and squares may be lit by an array of colors. Even if there is no evidence of other forms of synesthesia, are number forms really a special numbers-evoke-positions synesthesia?

We are not sure how common synesthesia is in the human population, partly because many synesthetes may not have realized that they are indeed different from other people. Also, it has been claimed that synesthesia is more common in children, but is gradually lost with age—and one can imagine that children are even less likely than adults to realize their unusualness, and certainly less likely to be believed. Current estimates are that one in a thousand women are synesthetes, but the frequency in men is up to eight times lower. So here once again we have a very clear-cut difference between the sensory worlds of the two sexes (although this sex difference is notably lacking in the prevalence of number forms). There is also very strong evidence that synesthesia runs in families, suggesting that it may have a strong genetic component.

One thing that is very clear about synesthetes is that they are attracted to artistic and creative careers, and this supports the idea that a major underpinning force in art is the search for links between different senses. It is often claimed that many of the great artists of history were synesthetes, but few of them left written evidence of this, and even fewer underwent specific tests for synesthesia. Listeners can see in their mind's eye the scenes that Vivaldi and Debussy were trying to evoke, although these composers left us written guides as to when we are supposed to hear the fogs of winter and the waves of the sea. It is impossible to read Dylan Thomas's *Miscellany* or Henry Miller's *Colossus of Maroussi* without being washed by a sea of merged sensations. And the colored lines and squares of Piet Mondrian's *Broadway Boogie-Woogie* really do pulse with a beat. But were these artists really experiencing synesthesia, or were they simply skilled enough to be able to elicit it in a nonsynesthete audience? Maybe that was where their genius lay.

Other artists have written of synesthesia as if they really experienced it. The painter James McNeill Whistler wrote, "As music is the poetry of sound, so is painting the poetry of sight, and the subject-matter has nothing to do with harmony of sound or colour." However, all the writing in the world does not prove synesthesia. Certainly, the members of Pink Floyd have subsequently admitted that the following rambling publicity material was at best drug-induced and at worst jokily constructed to sound hip: "Music in Colour by The Pink Floyd: For the first time outside London the Pink Floyd develop a Total Show and Explosive Sound with Films, Light and Image all combined into a single image."

The artist most often claimed to be a synesthete is Wassily Kandinsky. He often wrote about his attempts to unify color and music, and certainly his paintings have a remarkable sense of movement and rhythm. But does that make him a synesthete? As we have seen, a characteristic feature of synesthetes is that they do not consider their synesthesia to be unusual. To them it is unremarkable—it is simply the way the world is. While it may well inform their artistic work and cause them to make novel links, they are unlikely to consider it a significant conceptual basis on which to base their artistic career. Strangely, the most convincing creative synesthete was not an artist or composer, it was the American physicist Richard Feynman. In my opinion, Feynman was the greatest scientific genius of the twentieth century—making pivotal, incisive contributions to several different areas of physics— and unusually he was also a truly brilliant teacher. Feynman reported that when he looked at a mathematical equation, he saw the symbols in color and that these colors helped him decide what to do next with the equation. Yet he did not publicize this unduly, and he will never be remembered as the man who tried to introduce color into mathematics. He did not seem to consider his ability to be unusual, and that is the sign of the true synesthete.

So what exactly is going on in the brain of the synesthete, and can it tell us anything about sensation in the rest of us? It is difficult to know what implications synesthesia may have for the rest of us when we do not truly know how common the condition is. Many people may be constantly undergoing processes that I would consider synesthetic if only I could experience them. For example, are you sure you handle numbers in the same way as your best friend? There is some scientific evidence to guide us in the world of synesthesia, but it is scanty.

First of all, we know that the sensory organs themselves are not completely specialized to their tasks. All of them can react to things other than the stimulus that normally excites them. You can demonstrate this very clearly if you close your eye and press on it gently through your eyelid—as far to the side as you can, just in front of the bone. You will see a dull light flickering where you would normally expect to see the bridge of your own nose. If you move your finger up, the flicker moves down. If you move your finger down, the flicker moves up. This flicker occurs because the rods and cones in your retina are also sensitive to touch, so that if you distort the wall of the eye, they mistakenly tell your brain that they have seen light. You "see" the touch on the opposite side of the eye because the optics in your eye actually create a reversed image on your retina. So really, this lack of sensory selectivity is a low-level form of synesthesia common to us all. Maybe it is a reminder that our sensory cells all trace their collective ancestry back to simple, unspecialized ciliated cells.

Moving up the complexity scale, we have already seen instances in our evolutionary history when we have created two new senses from one old one, such as when our movement-detecting inner ear split into balance and sound-detecting parts. Conversely, we have seen that different senses can merge into one unified perception, as when we combine taste, mouth-feel, toxin-detection, and smell to create the symphony of sensation that is flavor. And at a more technical level, we have seen how many senses are combined and put in context in the tectum of the midbrain—hearing, vision, touch, and smell in most animals, infrared in snakes, and ultrasound in bats. In addition, we have seen that blind people who use sound echoes to detect nearby objects—so-called facial hearing—"feel" the resulting information as gentle pressure on their face. So when you look at it like that, we are all synesthetic beings.

But what can science tell us about the people who have gone a little further than the rest of us in their sensory union? There are a few discoveries that can guide us. Those of you who have misbehaved chemically at some time in your life may already be wondering how all this relates to the synesthetic experiences you underwent after indulging in lysergic acid diethylamide, or LSD. Well, acid-induced hallucinations may include synesthetic elements, but there are some fundamental differences between these and spontaneous synesthesia—they are, for example, unpredictable, random, inconsistent, and occasionally distressing. This in itself is probably evidence enough that drug-induced and natural synesthesia are too dissimilar to be comparable.

Other aspects of synesthesia appear more promising. For example, it is widely reported that if a synesthete for whom speaking evokes color is blinded in an accident, she retains her visual evocations even though she can see nothing else. Furthermore, these cases suggest that the seat of synesthesia is quite far along in the information-analysis process—somewhere up in the cerebral cortex, maybe. Even more remarkable than this, however, is the fact that in some color-blind synesthetes, spoken or written words evoke colors that they have never been able to see in the real world because of deficiencies in the photoreceptors in their eyes. They may give these colors strange names, such as "Martian colors," because they associate them not with what they actually see, but with their other form of visual perception. Thus, synesthesia can work on brain centers that have never had a chance to interpret information from real things in the outside world.

There is a further side of synesthesia that suggests possible avenues for study. You will remember that I said that an individual synesthete's responses to different words or sounds was entirely consistent. Well, different individual synesthetes differ in the sensations they evoke—there is no common code to convert words or characters to colors. However, there are some associations that crop up again and again. For example, for a large proportion of synesthetes the letter "R" or words starting with "R" evoke red. Also, for many, "Y" evokes yellow. You may now be wondering if all this synesthesia is based on a crude relationship between initial letters and colors, but that suspicion is thwarted by the finding that, among an English-speaking sample, the most common relationship of all is that the letter "D" evokes the color brown. Why should that be? And why do so many different people make this unexpected association? All synesthetes are different, but the fact that there are these striking commonalities suggests that there is an important underlying process.

At this stage it is worth mentioning a fundamental dichotomy in the synesthetic population that may help us in the future. Just as you or I can imagine hearing or reading a word, so can synesthetes. But do they evoke the same sensations when they imagine sensing something as when they actually do sense it? The answer to this is that some do and some do not. Some see the flicker of color when they imagine the word and some do not, and presumably they have accepted as normal whichever of these happens to be the case. It is hard to be precise at present, but does this not suggest that some people's synesthesia occurs at a different level in the sensory process than others? Some can obviously feed imaginary sensations in and get the evoked sensation out, and some cannot. Maybe it depends where in the sensory process the synesthesia occurs.

So what explanations do we have for synesthesia? All three theories are based on the idea that sensory information is meant to arrive at the brain in discrete streams that are then subdivided into substreams which can be processed in parallel. For example, you can imagine how visual information can be dissected into color, brightness, edges, and movement, and how these streams can be further dissected and analyzed. This subdivision of the streams is well underway as information passes through the thalamus, but we believe that it continues apace in the cerebral hemispheres. Yet at some point, all these different streams must flow together to give us a conscious perception of what is actually going on in the world outside—just as our optic tectum gathers different senses to produce an unconscious view of that world.

The simplest mechanistic theory for synesthesia suggests that abnormal links develop between different streams—for example, the text-interpretation stream spills over into the color-recognition stream. A second theory proposes that these between-stream links are a normal occurrence in infants, that we are supposed to lose them as we grow up, but that synesthetes do not. I must admit that I have a problem with this theory in that I personally cannot recall a vibrantly synesthetic childhood, and I am sure that I would remember something like that. Or would I? Anyway, the third theory is rather different. We know that higher stages in the pathway of sensory processing can exert profound effects on lower centers—the brain stem can dampen the response of the ear to sound, for example, and extreme stress can stop us feeling pain. We think that back-control of sensory pathways by higher centers is an essential part of how we perceive the world, altering our sensitivity to the world around us so we can ignore the minutiae and the obvious. This third theory suggests that at some point after all our sensory streams have started to flow together, signals can still be sent back down the chain to control our sensations. In synesthetes this system may be overactive, and one type of sensation can reach back and inappropriately evoke a different one.

Whatever the neuronal pathways involved, synesthesia is also challenging from the evolutionary point of view as it is reassuringly human in many ways. It most commonly involves interactions between two aspects of our perception that have undergone the greatest change in recent human history: language and color. Obviously spoken and written language are peculiarly well developed in humans, but as I discussed in my last book, color perception is also in an unusual state of evolutionary flux in our species. For example, we are one of the few mammalian species that can discern red, blue, and green, and yet our genetics are arranged so that a significant proportion of the human population (perhaps 8 percent of men) lacks this fine color discrimination. All this points to the possibility that language and color are new, rough and ready systems to us, still very much in the experimental stage. Is synesthesia a sign that our sensory system is having trouble coping with all this new linguistic and chromatic information? Maybe synesthesia is a hidden side of something we know that humans are inherently attracted to—metaphor. Are we like children entering a new playground of babble and hue?

The middle part of this book really has been an assault on the senses. Unified by the archaic little ciliated cells that gave rise to them, the different sense organs have driven the evolution of our brains. And we have seen how all the disparate movement and visual and chemical signals once detected by the descendants of those sensitive ancestral cells end up recombined in the higher centers of our modern brain. If the first part of this book is about the brain as geography, then this middle one is about the brain as archaeology—studying the ancient structures we have modified for new uses. Into the bargain, we have continued our epic journey onward and upward through the hindbrain, midbrain, and into the forebrain. We have also been figuratively edging our way across Galen's bridge from sensation to action. Now great things await us. With much of the wiring and plumbing behind us, we have just one region left—the endbrain: a hugely profuse radiation of mental power where reside our highest actions, perceptions, and hopes, and maybe even our consciousness.

15

INTERLUDE

Shrapnel and Magnets

It is amazing what I will do for my friends. I will fill out a question-naire about intimate piercings. I will spend over two hours in the grip of an enormous magnet. I will even—and this is the most humiliating bit—travel to south London.

Nick and I met at school when he was seven and I was eight. He was a strange little kid, but then again, I probably was, too. We were great friends all through school, and we still are. I liked the way he always had a different view on the world from everyone else, and how he always seemed to be a bit ahead of me intellectually. Or was it slightly to one side? I could never tell. Dr. Nick Medford is now a re-search neuropsychiatrist at the Institute of Neurology in Denmark Hill in London doing research on brain imaging in mental illness. I had already volunteered to help out as one of his supposedly normal control subjects, so it was no surprise when the call finally came, and Nick said he had wangled a slot on the MRI scanner, and could I please come over?

It was odd to be a subject in the experiment of someone I know, but it did at least mean that I was bought lunch first. Then I had to fill out a questionnaire that seemed to revolve around the two issues of whether I minded being in cramped spaces and whether I, or anyone else, had inserted any fragments of metal into my body. Although I feel I have lived a full life so far, I have not to my knowledge been showered by shrapnel in a combat situation, nor have I seen the need to sully parts of my anatomy with nuts and bolts. I knew that MRI involved a giant magnet, so I realized that this was the reason for the concern about metal paraphernalia. Presumably it is unpleasant to have the shred of shell casing in one's eye or the ring through one's scrotum grasped firmly by one of these monster magnets and moved forcibly several inches to one side. I am not claustrophobic, either.

Obviously, my familiar tormentor was not allowed to tell me what the experiment was about, but he did promise that I could have a copy of my pictures afterward—one of these hunky internal portraits is reproduced in this book (see Figure 4.7). I was asked to lie on a narrow metal bed with my head steadied on a brace and was then slid into the bowels of the MRI machine, which looked just like a giant white doughnut balanced on its edge. For the next couple of hours, the doughnut made some impressive buzzing noises and strange pictures were projected onto the screen in front of my eyes. And then it was all over. I found it unexpectedly relaxing.

Magnetic resonance imaging, or MRI, is the hot new way to look inside our bodies. Added to radiography and ultrasound, it gives us a formidable battery of imaging techniques. Yet no matter how many times I read about the theory of MRI, and although I understand it, I still cannot believe that anyone actually ever got an MRI machine to work other than in their own imagination. Magnetic resonance was first used to study atomic nuclei toward the end of the Second World War, but it was not until the 1970s that it was first applied to biological specimens—a tube of blood cells in 1973 and a person in 1977. Now there are thousands of MRI machines around the world, and we vets even make use of little ones zipping around in the backs of trucks.

The first stage in MRI is switching on the magnet. This is an enormous thing, and takes up much of the big white doughnut that encircled me. The magnet immediately got to work by aligning all the hydrogen nuclei in my head in the same direction. It does tweak a few other types of atomic nucleus as well, but these are less important. Hydrogen nuclei consist of just a single proton, which the physicists tell us possesses a quantum mechanical spin. This is why it can be lined up like a minuscule iron filing by magnets. I have a lot of hydro-gen in my head—all the H_2O of course, but also many of the other in-gredients of my brain, skull, and face. Much of the brain is hydrogen-rich fat, for example—the fatty insulating sheaths around many of those neurons, for one thing. Also, there is the strand of lard that shines with such embarrassing whiteness at the nape of my neck in the scan. Anyway, I am sorry to report that I could not actually tell when all my protons were being aligned. I was rather disappointed—I had expected my hair to stand on end at least.

Once my head protons were all obediently spinning about the same axis, zaps of radio-wave energy were focused on tiny points inside my head. These were tuned so that they would knock my protons out of the alignment induced by the magnet. Again, I could not feel any of this, but it does not seem to have done any lasting damage. Once knocked askew, the protons then fell back into alignment under the influence of the magnet, but as they did this, from their sheltered po-sition inside my head they radiated a burst of energy that was de-tected by a wire coil in the MRI machine. A computer then extracted all the different signal components in that single, noisy burst to work out exactly what the protons were doing at that little point inside my head. Repeated millions upon millions of times, the computer started to assemble the points, *et voilà!* A pointillist picture of the inside of my head. It all sounds too good to be true. Who would have thought that the quantum mechanics of hydrogen could be used to cut slices through me? But that is how it is done, and no doubt many of you will have had the same thing done to you, or will do in the future.

Between the buzzes, the screen flicked pictures before my eyes. The design of the experiment was not especially secretive. In an appar-ently random sequence I was shown images that were either boring (kettles, chairs, cars) or emotionally striking (car crashes, serious in-juries, screaming children)—not unlike watching a European art-house movie. Eventually it was all over, Nick slid me out from the maw of the giant doughnut, and we went for coffee. The idea of the experiment was clear—I was supposed to react to the emotive stimuli in a different way to the mundane ones. He was later to tell me that my strongest reaction of all was to a picture of Mickey and Minnie Mouse. Filth. Absolute filth.

I was a supposedly normal control subject for a study of depersonalization disorder, an uncommon condition of which many of us have nonetheless had a taste. We do not really know how many of us experience occasional brief episodes of depersonalization during our lifetimes. I think I did it once as a teenager, at the drowsy boundary between sleep and wakefulness. It must have been around noon. You get a strange feeling of detachment from your own body and your surroundings. You know you are you, but you feel that you are somehow removed—watching yourself as if in a movie. Your sensations, feelings, and actions are separated from you. You feel numbed. And then everything slots back into place and you are inside yourself again.

Some of you may recognize this feeling, but for sufferers of depersonalization disorder, it is a constant, or at least recurrent, experience. It often starts in the late teenage years, sometimes after an episode of stress, depression, or drug use. It may even be caused by emotional trauma from earlier in childhood. Whatever the cause, the nature of depersonalization disorder seems consistent—it is unremitting and permanent. These people know who they are even though they feel detached, and most of their other faculties are normal. Although they cannot escape the feeling that they are watching themselves in a dream or a movie, they know this feeling is false. Because of this, a surprising number are able to adapt and learn to live with the disorder. Neither drugs nor psychotherapy seem to make much impact on this sense of unreality.

A common symptom reported by people with depersonalization disorder is that they feel emotionally numb. Although they are slightly unresponsive to the world around them, they are aware of what they are smelling and hearing and seeing, yet they do not seem to be able to ascribe much emotional significance to external things. It is as if the world is drifting irrelevantly by. It is there, but it has no real significance. There are rare cases around the world of people who, after accidents, lose their ability to ascribe emotional weight to the things and people around them. Some of them explain the lack of emotional feeling by constructing complex conspiracy theories in which their loved ones have been replaced by malign but accurate impostors. In contrast, depersonalization disorder is sufficiently common that it provides us with a reasonably sized cohort of people in which we can study emotion.

And it was this cohort with which I was being compared. I was actually undergoing a functional MRI, which allows scientists to watch activity in different parts of the brain as it happens. And Nick told me that emotive and nonemotive stimuli caused flickers of activity in entirely different parts of my brain, as he had expected. In particular, the crashes, grimacing, mutilation, and cartoon mice stimulated areas thought to be important in "disgust perception," including the insula, or "island," of Reil. We will find out what this island actually is very soon, but for now please remember that it is not stimulated by emotive images in people with depersonalization disorder. In them, the island of Reil was only activated by the more neutral pictures.

So we can now take snapshots of our brains and what is going on in them. We can even be spectators as we carry out the very highest levels of sensation and perception. All that dry sensory information is distilled down into emotion, pleasure, and disgust, and we can actually see it happening. We have observed the place in our heads that tells us who is important and whom we should care about—the island which ensures that no man is an island.

III

WHERE ALL THE MIND MAY BE FOUND?

The itinerary: A flight above the islands of Calleja and the radiations of Zuckerkandl

We have stumbled unexpectedly upon the location of the neuronal patterns
"which dreams are made of" and have glimpsed other mechanisms
within the humming loom of the mind.

—Wilder Penfield and Theodore Rasmussen *The Cerebral Cortex of Man*

16

THE BRAIN AS ENGINEERING

Wilder Penfield and the Cortex

The patient usually lay on his side, with the diseased hemisphere upward. The surgical site was surrounded by sterile drapes, although the patient's face was always left uncovered so he could communicate with his surgeon.

The operation was the culmination of a long process of clinical examination, medical tests, and the gaining of absolute trust. After all, performing the osteoplastic craniotomy, or "Montreal procedure" as it was sometimes known, on an entirely conscious patient was always going to require an unusual relationship between surgeon and patient. Doctor Penfield often developed strong bonds with those in his care, and this was crucial for the procedure to be successful. For the 1930s and 1940s, this surgery was remarkably brave and successful. Local anesthetic was injected into the scalp in a ring pattern, the skin incised, and a large slab sawn from the skull. This bony flap was folded back, and the dura mater, the outermost and toughest meninx, was incised and peeled away. The cerebral cortex beneath was sprayed with a fine sterile mist to keep it moist and the air flowing over it was irradiated with cleansing ultraviolet light. No ultraviolet was allowed to fall on the brain tissue, however, for fear of burning it. Penfield worked methodically and logically, placing little numbered sterile squares of paper onto the convoluted pink organ, trying to determine where lay the seat of disease, and where the cuts could be safely made. All this with the patient conscious, seeing, hearing, and speaking freely with the man who now was dabbling his fingertips inside the patient's head.

Wilder Graves Penfield was born in Washington state in 1891, although he did all his greatest work in Canada—a kind of reverse Joni Mitchell, but with electrodes. Although he later wrote of his physical and mental ineptness in childhood, this was almost certainly excessive modesty, as he was to study at Princeton and become an accomplished wrestler. Rather than preparing him for the delicacy of his future career in neurosurgery, this latter pursuit apparently yielded superb neck muscles that I would certainly envy after a few hours in the operating theatre. He was determined to win a Rhodes Scholarship to Oxford, and when he failed at the first attempt, he spent a year raising money for his medical training by coaching the Princeton freshman football team. His second attempt was more successful and he was soon on his way to Oxford.

While in England, Penfield was inspired by the network of European scientists who were trying to piece together the functions of the different parts of the nervous system into an integrated whole. The anatomical geography of the brain was now known, and there was also considerable agreement about its evolutionary archaeology—the known and the unknowable. Now the focus of neuroscience had turned to how the brain functions as a machine—the potentially knowable. Brain as engineering, as it were. Penfield was enthused by the idea that the brain was the new frontier and used the explorer's idiom to describe it. Just as we compared the brain to the unknown periphery of a medieval map at the start of this book, when Penfield was at Oxford it was described in such terms as, "the undiscovered country in which the mystery of the mind of man might some day be explained." Scientists were moving away from Descartes' separation of mind and brain-machine toward the idea that the mind could instead be defined entirely in terms of the electrical activity going on in the corrugated mass inside our heads.

Fired by this idea that the mind was accessible to scientific investigation but also driven by a strong urge to help others, Penfield continued his medical studies at several centers in Europe before returning to his native United States, where he worked at the Johns Hopkins Hospital in Baltimore, the Peter Bent Brigham Hospital in Boston, and Columbia University in New York. From there he moved to Montreal, where he was to spend the rest of his career, in 1934 founding the Montreal Institute of Neurology with over a million dollars from the Rockefeller Foundation. Penfield had realized that by becoming a neurosurgeon he would gain a privileged access to the human brain that no lab-bound scientist could ever attain, all the while helping those in need. His empathy with patients was well known, and in no small part contributed to his ability to treat their distressing illnesses. In one particularly striking case, he performed a radical excision of a large region of cerebral cortex to remove a malignant tumor. It was impressive enough that, having discovered the alarmingly aggressive spread of the tumor during surgery, Penfield opted for a boldly extensive removal, which he correctly predicted would have no severe adverse effects on his patient. What is more dramatic is that he was able to exercise this degree of skill, insight, and sympathy while delving inside the skull of his own sister.

For such an important organ, the brain is remarkably vulnerable. It is one of the most metabolically active organs in the body and is constantly seething with electrical impulses whizzing along its billions of axons. It really does seem as if our brains have been engineered to the very limits of what is possible, and because of this we are especially prone to suffering failures of that engineering. Many of the patients who walked through the doors of Penfield's Institute were suffering from epilepsy—uncontrolled storms of activity in the brain that can cause a wide variety of effects. During a seizure, sufferers can experience a variety of strange sensations, feelings, and emotions, and the episodes are often preceded or pervaded by a confusing, altered mental state—an aura. Often, they lose consciousness and may undergo dramatic and violent movements.

The variety of things that cause seizures is testament to the brain's sensitivity. Many of Penfield's patients had a known underlying cause of their disease within the brain itself. Rather than any specific evidence of functional loss of any one brain region, the first sign of brain disease is often seizures. The brain, and especially the cortex, is often remarkably good at coping with the destruction of small sections of itself, but once that damage alters the brain's pattern of electrical activity, there seems to be no internal mechanism for controlling the resulting electrical storms. The delicate balance of activity in the brain is also often the first thing to fail when the body undergoes severe metabolic imbalance—fits often result from uncontrolled liver or kidney failure. Certainly other organs are affected by the chemical chaos caused by these conditions, but they seem to be better at withstanding the bad times than the sensitive, demanding brain.

Very often we simply do not know what causes seizures. In human and animal patients in which no within-brain or without-brain cause can be found, we must assume there is some focus of abnormal activity in the brain that triggers the attacks. Often Penfield would open the skull to find that the brain looked essentially normal—no ominous reddened lumps distorting the sculpted surface. Humans are not alone in suffering from this essentially mysterious or "idiopathic" epilepsy, but they certainly have a larger mass of writhing cortex in which it can start than the simpler souls I treat in veterinary practice. One thing we do know is that some individuals are inherently more prone to seizures than others, and this has led to the concept of "seizure threshold." For example, anesthetists have known for some time that some patients are more likely to have a seizure when administered certain drugs. People are pretty consistent in their seizure susceptibility, and this seems to be something that is largely controlled by the genes we inherit. So for seizures to be induced, some innate threshold must be exceeded, meaning that seizures are a combination of provoking stimuli and intrinsic electrical susceptibility. We all have a seizure threshold, even if it is never reached in our lifetime.

If seizures result from the unusual amount of activity going on inside what Penfield called the "humming loom of our minds," then some of the other common catastrophes of the cerebral cortex result from its sheer size. Not only is your brain at the limit of how much electrical processing can be shoehorned into an organ, it is also pushing the envelope of how big a brain can be. Some animals do have larger brains than humans, but only very few (mainly large mammals like whales and elephants). The size of the human forebrain has led to some engineering problems that most other species do not face. At the risk of oversimplification, if a dog is struck on the head, then that blow is either so mild that the animal is relatively unaffected, or it is so destructive that the animal dies. The bloated human brain behaves strangely on impact, and this leads to a complex syndrome called concussion.

The brain is by no means a rigid, impact-absorbing structure. It may float, cushioned in a sea of cerebrospinal fluid, and be covered by a hard, bony cranium, but it is itself only semisolid: a concoction of fleshy cell bodies and fatty axonal strands with little connective tissue to hold it all together. When you think about how the brain reacts to injury, it is helpful to view the brain this way—as a half-set pudding bobbing in a liquid sauce inside a rigid container. The head may be struck by sharp objects that neatly puncture the skull and lacerate the patch of brain beneath, but this simple form of brain injury is not particularly common. Instead, many more damaging injuries result from blunt trauma to the head that often does not even breach the skull. Around one-quarter of all human deaths due to trauma result from brain trauma, and three-quarters of these occur in males. That perhaps half occur in the inebriated or drug-incapacitated is evidence that the injuries are often crude and simple—the smack of the head on the sidewalk, the sudden jarring as the car strikes the tree.

As long as the skull is not fractured, the brain is actually rather good at surviving the one-off single, simple impact, and perhaps this is because this is the sort of injury we were most likely to suffer as we scampered about the ancient savannah. Repeated impacts are less well received, however. One problem is that the fluid in which the brain floats can itself cause unexpected problems. Obviously, its presence means that the brain is slightly cushioned from the impact the skull receives, but that cushioning itself requires that the brain is not very well anchored onto the inside of the skull. Apart from the points where the cranial nerves and spinal cord escape the cranium, the brain is a remarkably free organ. Free to drift and slide, but also free to bounce. Consider what happens when a head strikes a pavement. You may be surprised to know that the first part of the brain to be injured is not the side nearest the impact, but the opposite side. Your brain is actually slightly less dense than the cerebrospinal fluid, so it has less inertia than the fluid in which it floats. On impact, the heavier fluid flies toward the impact site, displacing the brain away so that it hits the opposite side of the skull. Thus, this initial "contrecoup" injury, facing away from the impact, is often the worst injury sustained—something that mystified neurologists for many years. Once this injury has occurred, the rebound of the head away from the pavement then flicks the denser cerebrospinal fluid in the opposite direction, so that the brain is displaced violently toward the impact site, where it may suffer a lesser "coup" injury. So the brain is often injured in a counterintuitive way because of its buoyancy within its bathing fluid. Why nature could not have equipped us with cerebrospinal fluid of exactly the same density as our brain is one of those irritating frustrations of biology.

There is one effect of trauma on our huge forebrain that no amount of careful juggling of liquid densities could prevent. Our brain is not a solid object, and in the split second of impact it can behave worryingly like a liquid. If you fill a bucket with water and then start to spin it, you will notice that the water around the edge rapidly speeds up to keep pace with the bucket, whereas the water at the center lags behind for some time—appearing stationary for a considerable period. Unfortunately, this same effect also occurs when a human head is violently rotated, such as when a car driver is suddenly decelerated to rest after hitting an immovable object. The brain tissue nearest the skull is dragged around by its rotating bony container, but the center of the brain stays where it is. If the rotation is rapid enough, the brain's core can effectively dislocate from its surface—the connecting axons sheared by the twisting movement of one part of the brain against another. This rotational shearing of the semisolid brain tissue may underlie many cases of concussion, whether mild and temporary or severe and fatal. Yet this effect is almost unknown in veterinary practice. Although animals do not drive cars, they are often hit by them, and these creatures often show other signs of violent head rotation—fractured jaws and neck vertebrae for example. However, they often do not show signs of humanlike concussion. This is almost certainly because their brains are simply too small to produce a suitably dramatic bucket-of-water effect. A big dog's brain is the size of a small woman's hand. Its center is simply not far enough from the skull to shear away from it.

As well as electrical overload and crude size, there is a third and

final engineering problem with the human brain—its complete imprisonment inside the skull. As I mentioned some time ago, it is para-
doxical that the organ designed to help us interact with the outside
world is the organ most completely sequestered from that world.
Obviously the brain is locked in a box for its own protection, but
this rigid confinement can also be its undoing. We have already seen
how hydrocephalus is more severe when it occurs in the rigid cranium of the adult rather than the flexible braincase of the infant, but
rigid confinement is also a problem in other situations. For example,
impacts can cause small blood vessels around the brain to burst, causing large blood blisters either outside or inside the dura mater—the
epidural and subdural hematomas I mentioned in Chapter 5. In the
confinement of the skull, these hematomas start to press on brain tissue and destroy it—largely because the brain tissue has nowhere to
go. The same thing can happen when the brain's blood vessels burst
unexpectedly as part of a stroke, although why human blood vessels
are so prone to this ridiculous weakness is a mystery—most other
mammals are entirely immune to strokes.

The risks posed by the brain's imprisonment in the rigid skull
would be bad enough were they only to come into play when blood
blisters form. However, when this imprisonment is coupled with the
fact that the brain has a ridiculously misjudged propensity to swell
when injured, you can see that the brain/brainbox combination is an
accident waiting to happen. As a veterinary student, I was surprised
to learn that most of the damage caused by injuries to animals' nervous systems is indirect. Obviously these injuries squash neurons and
sever axons, but they often also cause profound swelling of the nervous tissue, which in turn leads to greater damage and thus further
destructive swelling. When a dachshund fulfils its genetic destiny and
propels a slipped disc into its spinal cord, we pump it full of drugs to
reduce the swelling. And if that fails, we saw windows in its wonky
vertebrae to give the cord some freedom to bulge rather than actually
repair anything. And this perverse tendency to swell is inherent in
your spinal cord and brain, too.

Almost anything going wrong in a brain can make it swell—an impact, an infection, a tumor. And when it starts to swell, its options
are limited. It can displace a few teaspoonfuls of cerebrospinal fluid

out of the big hole (foramen magnum) at the bottom of the skull through which the spinal cord passes, but after that it is in trouble. As it slowly enlarges, the whole brain starts an inexorable destructive migration through that foramen magnum—it is partially extruded through the bottom of the skull like so much toothpaste from a tube. First to go are a couple of dangly bits at the bottom of the cerebellum, the tonsils, which are squashed as they are forced into the foramen magnum, a process of distortion known as "coning." Farther forward in the skull is a tough, taut curtain of dura, the tentorium cerebelli, or "tent of the cerebellum," which partitions the cerebellum from the cerebral cortex. As the cortex swells, its hindmost part is squeezed under this curtain into space usually occupied by the cerebellum—this tentorial herniation not only damages the errant cortex, but also forces it against the brain stem, causing irreversible damage to that, too. To make matters worse, there is also a larger, crescent-shaped curtain between the two cerebral hemispheres, the falx cerebri or "sickle of the brain," so if one hemisphere swells faster than the other, its deepest portions are thrust under the falx to the opposite side.

As you can see, the cerebrum can pay a terrible price for being so active, huge, and enclosed. We evolved our massive forebrain in quieter times, and it struggles to cope in these days of rampant diabetes, car accidents, and baby-shaking nannies. Wilder Penfield was working at a time when these cerebrum-unfriendly influences were starting to come to the fore, but he also realized that the cerebral cortex has one significant engineering feature which we can use to our advantage. You will remember that I pointed out in Chapter 3 how well the brain copes with the day-to-day insults of mild trauma, intoxication, and general wear and tear. Well, despite all its failings, the cerebral cortex is indeed remarkably adaptable in some ways. One bit of cortex looks pretty much like any other, and this homogeneity is a sign that if one region of cortex is damaged, then another region just might be able to assume its role. There seems to be considerable spare capacity and a good deal of flexibility up there. Regions of cortex can help each other out in a way that is simply not possible in the hard-wired world of the brain stem. This adaptability is one reason why Penfield could take out approximately a quarter of his sister's cerebral cortex with surprisingly little effect. It also explains why whole hemi-spheres can be excised from epileptic infants, yet these children grow up reasonably normally.

Most of Penfield's patients had epilepsy—they had not suffered contrecoup injuries, twisting concussion, or brain swelling—so all he had to do was work out what he could and could not chop out. Hence all the little numbered squares of paper. Over half of his epilepsy pa-tients were cured by his surgery, so Penfield must have been doing something right. Sometimes he could see what was wrong with the brains exposed in front of him, and sometimes he could not. Either way, he tested every exposed area of cortex with electrodes buzzing with alternating current at up to seventy volts and sixty cycles per second. If he stimulated an unseen tumor, he could often detect it by the abnormal pattern of electrical activity it emitted. But remarkably, if the focus of epilepsy was a visually normal patch of brain, he found that he could still identify it because when it was electrically stimu-lated, the patient would report that they were experiencing some-thing very similar to their usual epileptic auras. Penfield had suc-ceeded in his proclaimed intention to be "a neurologist in action"—he had located the seat of a natural brain disease by activating the conscious cortex of another human being.

However, the next step in Penfield's meticulous process was, if any-thing, even more dramatic. He now proceeded to use his electrodes and the verbal commentary of his trusting flip-top patient to discern the function of all the surrounding areas of cortex. This way he knew what handicap he might inflict on his patients by overenthusiastic slicing. When he stimulated certain regions, he obtained specific, re-peatable responses—consistent in the individual and also extremely similar among patients. Here he was, fulfilling the dream of anatomy-obsessed natural philosophers throughout the ages, mapping the roles of different areas of living human brains. Five millennia after Imhotep first described the convoluted surface of the *ais*, Penfield was charting its invisible functional geography. And that geography seemed encouragingly constant throughout all his subjects. The exact details might differ, but he found that stimulation of a given region al-ways caused generally similar responses. And in some regions the ex-act details did not differ at all—the brains of different people had

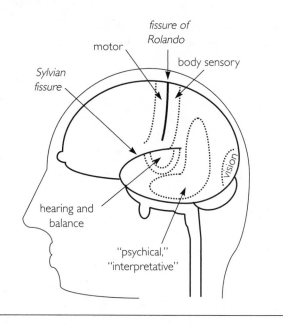

Figure 16.1. The cerebral hemispheres—a simplified version of Wilder Penfield's maps of the brain, derived from information from operations in which he electrically stimulated the cortex of conscious humans.

identical functions in identical places. It was a wonder of anatomy—like the old phrenology, but this time it was the truth.

So what effects did his electrodes elicit in his patients? Reading the reports of Penfield and his collaborator Theodore Rasmussen, it is easy to see that they were overwhelmed by excitement at their findings. Although the disinterested reporting of results in scientific writing is supposed to be kept separate from the interpretation of those results, the reader often has the feeling that these authors simply cannot wait to tell you the importance of what they have found. I doubt that today's science journals would have much truck with scientists who claimed to have detected "the neuronal patterns 'which dreams are made of.'"

Much of the action was centered around the fissure of Rolando, a prominent vertical groove that slices each hemisphere into front and back halves (Figure 16.1). Penfield discovered that this cleft is a visual marker separating the two regions that gave him his most repeatable results. Actually, the presence of the fissure is entirely fortuitous, as most mammals do not have a cleft in that position at all. When he stimulated the strip of cortex behind the fissure, the patients told the experimenters that they felt a tingling or numbness in a certain part of their body—always on the other side of their body, in fact. In usual Penfieldian fashion, each bit of cortex always induced the feeling in the same part of the body, and so consistent was this relationship that Penfield was able to draw a map of the body over his diagram of the cortex (Figure 16.2). He had identified what we now call the primary sensory cortex: the slab of cerebrum that first receives tactile information from the body—the terminus of those ribbonlike lemnisci which we last encountered snaking their way up through the brain stem.

Conversely, Penfield found that the strip immediately in front of the fissure of Rolando contained a motor map of the body. Here stimulation induced involuntary trembling or paralysis of specific parts of the opposite side of the body (Figure 16.2, right side). This strip is now called the primary motor cortex. It is where final movement commands are dispatched down, via the spinal cord, to the muscles of the body—it is the starting point of those pyramids that course down through the brain stem. Penfield had located the far end of Galen's bridge of sensation, perception, interpretation, motivation, planning, and action. And when he noticed that the primary sensory and motor cortices are connected to the opposite side of the body, Penfield was confirming what we already know, that the forebrain is wired back to front. Neuroanatomists had known for some time that the lemnisci and pyramids switch sides on their long journey between body and cortex. They "decussate."

The fact that Penfield could account for all the motor and tactile functions in two little strips of cortex was tantalizing in the extreme. He already had a great deal of function under his belt, but he had a lot of cortex still unmapped. Next, he discovered the primary visual cortex—the little patch at the very back of the hemispheres where he could make his patients see lights, forms, and movement on the opposite side of the body (Figure 16.1). Hot on its heels came the primary auditory cortex, where lay buzzing, humming, ringing, and hissing, and the primary vestibular cortex: the seat of dizziness and vertigo. And still most of the cortex remained uncharted. Could the rest of the mind be so remarkably amenable to decipherment?

As he crept away from these primary cortices, the maps became

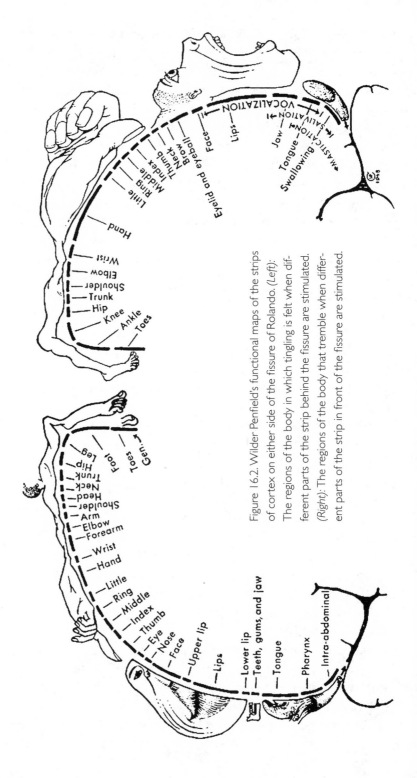

Figure 16.2. Wilder Penfield's functional maps of the strips of cortex on either side of the fissure of Rolando. (*Left*): The regions of the body in which tingling is felt when different parts of the strip behind the fissure are stimulated. (*Right*): The regions of the body that tremble when different parts of the strip in front of the fissure are stimulated.

vaguer and less predictable but even more interesting. As the electrode wandered from the visual cortex, for example, different visions appeared that were no longer restricted to the opposite side. It seemed as if the eminent doctor was moving to regions where the crude, basic sensations are analyzed, combined, and interpreted. And a sideways shift from the motor cortex led to more complex movements—stiffening of entire limbs, for example, and most remarkably of all, vocalizations. Greater excursions from the certainty of the primary cortices yielded yet more intriguing effects. Stimulation of particular, well-defined areas caused specific and complex abnormalities in the construction and articulation of spoken language, and as we will see, we today believe that these areas represent the physical location of our linguistic abilities.

Perhaps the most exciting discoveries came in the wide-open spaces on the side and top of the brain. Here Penfield induced effects that he could only describe as "psychical" (Figure 16.1). All the sensation in the world is useless if it cannot be understood, interpreted, and placed in context, and remarkably Penfield succeeded in locating the areas of cortex that do exactly this. He classified the diverse psychical effects he observed into two groups, the first of which he called "experiential hallucinations." Sometimes his patients would report vague feelings, such as being somewhere else, or being far away, or feeling rather silly. At other times, the electrode's touch would evoke complex reminiscences of the past. Surprisingly often, these were memories that his patients had not recalled for some time, and which they believed they had remembered only after being "reminded." The hallucinations were often extremely rich in texture and feeling, and they occasionally gave wonderful insights into the mind. Patients might recall episodes they had not themselves undergone, but had instead experienced vicariously by reading a book or childhood comic. They even reminisced about experiences they knew full well had occurred only in a dream. They were not passive observers, but active participants—they recalled not just childhood frights, but also the feel of their own pounding heart and their fleeing legs.

Penfield's second class of psychical effects were perhaps related to the stage after which sensations are compared to memories of previous experiences. Instead, these "interpretative illusions" appeared to arise from the regions of brain that finally apply a context to incoming sensations. Astoundingly, stimulation of some areas of the side of the brain could induce feelings of context or interpretation without any actual associated sensation or reminiscence. At the simpler end of the range, the electrode might induce altered perceptions of distance or intensity of things in the real world. It could also elicit feelings of strangeness, absurdity, fear, or disgust—one is reminded of my response to Mickey and Minnie. The alternating current could even cause patients to reassess their status in their environment, inducing *déjà vu* and *jamais vu*—phrases from Penfield's adopted bilingual city that mean so much more than just disquietingly explicit feelings of unexpected familiarity or unfamiliarity.

It was almost frightening how far Penfield had come in his search for the workings of the higher brain. He laid the foundations for much of modern neuroscience because he showed us that, just as the old anatomists had always hoped, many things in the brain do have a defined location. The mind, it seems, has a definite sense of place. Penfield strode boldly in "the undiscovered country in which the mystery of the mind of man might some day be explained . . . I am an explorer, but unlike my predecessors who used compasses and canoes to discover unknown lands, I used a scalpel and a small electrode to explore and map the human brain." A place for everything and everything in its place. Some of Penfield's findings, such as his perhaps over-strict localization of some functions, have been reassessed since his pioneering work, but his essential message has outlived him. There are places in the brain where all the mind may be found, if only we know how to look. In the remaining chapters we will hunt for some of those places, but we will keep our hunting short and to the point. We must be especially wary of the deceitful specter of consciousness—that which we all experience but cannot define. We are, after all, moving into unfamiliar territory, but do not be unnerved. Let us dance lightly through the fields of the mind.

17

THE APPARENT DISORDER
OF THE CEREBRAL JUNGLE

What Is in Those Hemispheres?

Long before Penfield it was obvious that the endbrains were important. The cerebral hemispheres are, of course, enormous—entirely dominating the rest of the brain—and especially huge in supposedly intelligent humans like us. Their surface is also often satisfyingly convoluted, again particularly in humans. It was always tempting to ascribe the highest achievements of the human intellect to the hemispheres. And when neuroscientists bit the bullet and started experimenting on them, their suspicions were confirmed. As the nineteenth-century savant Ernst Haeckel gleefully reported, "It is possible to remove the great hemispheres of a mammal, piece by piece, without killing the animal, thus proving that the higher mental activities, consciousness and thought, conscious volition and sensation, may be destroyed one by one, and finally entirely annihilated." Haeckel did not actually carry out this experiment himself, but we must imagine that it was a pretty gruesome exercise. So the hemispheres are undoubtedly important, but what are they?

Back in Chapter 4 we saw that the two endbrains bulge out, one on each side of the forebrain (see Figure 4.5b). They seem like an exceptional elaboration of what is otherwise a linear nervous system, and we suspect that this may reflect their status as a relatively recent addition. All jawed vertebrates have them, but lampreys do not, although they are such vilely unusual little creatures that I suspect they could have discarded their endbrains deliberately to confuse evolutionary biologists. Jawless hagfish, in contrast, have something that looks very like endbrains. Anyway, the embryonic development of the endbrains certainly has the air of an afterthought about it. Unlike the rest of the nervous system, which is induced to form by the underlying notochord, the endbrains probably grow under the influence of the nearby mouth and throat. Maybe the endbrains evolved as we made the transition from bumbling, passive filter feeders to more actively mouthy food grazers and stalkers.

The general arrangement of the human cerebral hemispheres is really very unlike that of the other parts of the central nervous system through which we have traveled. The spinal cord and brain stem consist of grey-matter nuclei and horns surrounded by the white-matter axon bundles. In the hemispheres, however, the grey matter—the cell bodies themselves—are on the outside and their white-matter interconnections are on the inside. This inverted arrangement of grey and white is dramatically at odds with most of the rest of the brain. Instead of scattered blobs of grey, the grey in the cerebral hemispheres is arranged in an extensive, continuous sheet that covers the surface. This coating of grey is the cortex of which we hear so much—the word means "rind" or "bark." The cortex is never more than a quarter of an inch thick, so each half of it can be visualized as an irregularly shaped, two-dimensional sheet of interconnecting nerve cells. It may seem strange after all the three-dimensional shenanigans we have been through to be presented with a simple sheet as the supposed pinnacle of cerebral development, but there you have it.

This two-dimensional arrangement is actually quite a novel development. The endbrains exhibit greater variation among the different groups of vertebrates than almost any other part of the brain, and the mammalian configuration is by no means typical (Figure 17.1). We can elucidate the story of the evolution of our cortex by comparison with the other vertebrates who have survived to the present day. The original layout of the endbrains was thought to be two symmetrical lobes, each containing a fluid-filled ventricular space at the center. This ventricle was surrounded by five different regions—inner, upper, outer, basal, and "septum"—but these regions were not demarcated into an outer zone of grey matter and an inner zone of white. Because of this, evolutionary biologists do not use the term "cortex," but instead call this tissue the pallium, or "cloak," as it enwraps the ventricle within.

These five chunks of pallium have fared differently as the vertebrates have evolved, and for reasons we do not understand, they have had radically different fates in different groups. Sharks and other cartilaginous fish have inherited the original basic plan, as did the amphibians, which gave rise to all the land vertebrates. There were a few variations on the theme—especially in sharks, which appear unusually "cerebral"—but the system is essentially the same. In the bony fish all sorts of unexpected things happened. First of all, the pallium almost split open and unfurled into a strange shape resembling a pair of ram's horns. Then the internal subdivisions among the different regions blurred, and the pallium merged into one unified mass.

The vertebrates that were able to spend all their lives on land—the reptiles, birds, and mammals—all developed one part of the pallium at the expense of the rest, presumably to deal with the additional sensory and motor skills required to live in this harsh new environment. In reptiles the outer region of the pallium is greatly enlarged, and it bulges far into the ventricle as the dorsal ventricular ridge. In birds, who are more closely related to lizards and their kin than we are, the outer region is even more greatly enlarged, almost obliterating the ventricle, and is called the wulst, a German word meaning "bulge." The wulst is probably a processing center for the enormous amount of visual and balance information that is part and parcel of birdy life. Both the reptilian dorsal ventricular ridge and avian wulst also receive diverse inputs from throughout the nervous system, and so are probably the functional equivalents of our cortex. Thus, this may be where these fellows do most of their thinking, whatever form that takes.

In mammals the endbrains evolved along a different path. Not only did we concentrate all our grey matter into that outer layer of cortex, but it was instead the upper pallium that yielded most of that cortex. The side region formed the small piriform—and it really is "pear-shaped"—lobe on the underside of the brain, whereas the inner regions form a relatively small but extremely important structure, the

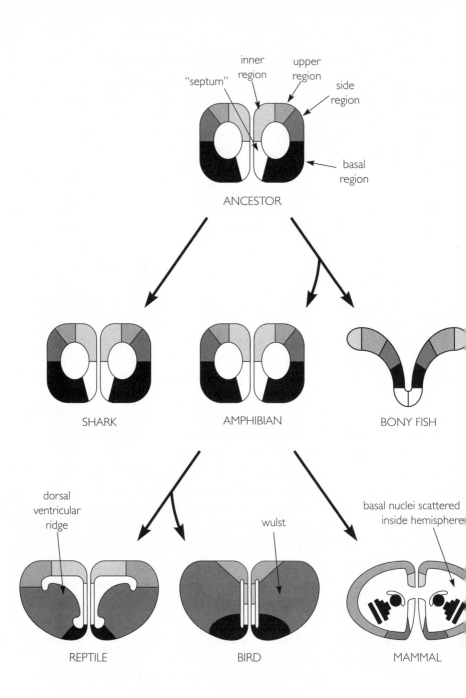

limbic system, to which we will return. And we think that the basal region of the pallium has done something very unexpected in mammals (see Figure 17.1). It seems to have detached from the rest of the cortex, crept inside it, and fragmented to form the basal nuclei, isolated islands of grey matter. In our discussions of Parkinson disease, we saw that these islands are involved in the control of movement.

So the mammalian brain is quite a departure from the standard vertebrate plan, but for anyone trying to cite this as evidence of our higher status in the scheme of things, it must be admitted that it is bony fish that have altered their pallia the most. Brain anatomy has often been used to support the idea of some animal groups being more "advanced" than others, and indeed some people have used it in the same way for human races. Whatever the politics and ethics of the latter, to a modern-day evolutionary biologist the idea of some groups being "advanced" and others "primitive" is ridiculous. Some species may have deviated from the ancestral scheme less than others, but if they have survived successfully to the present day, then who is to say that they are somehow inferior? This is a hearteningly even-handed

---

Figure 17.1. A theory for the evolution of the vertebrate endbrains, or pallium. Each stage is represented as a cross section through the endbrains. In the ancestor of the jawed vertebrates, the two endbrain swellings each consisted of a central cavity or ventricle surrounded by five main regions. These five main regions have been expanded and modified to different extents in the various vertebrate groups. Sharks and amphibians retain the original basic plan, but the endbrains of bony fish have become extremely distorted—virtually turning inside out in the process. In addition, the different regions of the bony fish brain have blurred into one homogenous mass. In reptiles the side region of the ancestral endbrain has become predominant, and it bulges prominently into the ventricle as the "dorsal ventricular ridge." In birds this arrangement has become more extreme, and the large side region, or "wulst," has almost obliterated the ventricle. In mammals, it is the upper region that predominates, forming most of the mass of the cerebral cortex. The inner region forms the hippocampus and associated structures and the side region forms the piriform cortex, involved largely in the sense of smell. The basal region has migrated away from the outside of the brain and now lies embedded deep within it as the basal nuclei. Also, an additional link has been formed between the two sides—the corpus callosum.

way of looking at things, but it is still remarkable how often the words "primitive" and "advanced" crop up in scientific books.

This old mammal-centric chauvinism was written into the names we used to give the different parts of the cerebral hemispheres. We used to call them the paleo-, archi-, and neocortex (ancient, old, and new) on the assumption that we acquired them in that order, sequentially adding new regions in our inexorable ascent to mammalian supremacy. Many theories about brain evolution were based on this concept, including the idea that there was something about the more recently evolved regions that rendered them more vulnerable to disease or degeneration. Yet our more recent realization that all vertebrates have the same basic components to their endbrains and that mammals just happen to have concentrated on one of them—the upper pallium—has rather put paid to this concept of old and new. There is nothing inherently modern about our mammalian cortex. All vertebrates have an equivalent region—they simply do not get so obsessed by it.

Another commonly quoted mammal-centrism is the idea that the cerebrum of nonmammals is really little more than a center for processing smell sensations, and that it is only in mammals that it has acquired its status as a repository for our higher functions. This is of course very reassuring for furry lactating types, as it gives us a sense of superiority, even perhaps suggesting that we are the only animals that can truly think or feel. This preconception also perfectly supports our aesthetically pleasing idea that each of the three brain bulges originally evolved under the influence of one of the three major senses—hearing and balance for the hindbrain, vision for the midbrain, and smell for the forebrain. Yet when we look a little closer, we see that the supposedly mammalian tendency to concentrate higher processing in the forebrain is actually evident throughout the vertebrates. In fact, there is no living vertebrate in which the cerebrum does not receive inputs from just about every other part of the nervous system—no living vertebrate in which it is just a "smell-brain." To a greater or lesser extent, the forebrain always acts as an integrating center for sensation and action. There seems to be something inevitable about the drive to push our more complex processes up there: to "cephalize."

So if our "higher functions" are up in the forebrain due to some enigmatically potent evolutionary drive, why do we also do our smell processing there? Well, the answer to that is probably altogether more prosaic: your nose is at the front of your head, so it plugs into the front part of your brain. Your eyes are next, so they get the midbrain, and your ears are behind them, so they get the hindbrain. Nature is often sweetly parsimonious like that. So the forebrain has to house the unlikely bedfellows of our "higher centers" and our sense of smell, and these two things are probably there for entirely different reasons. Yet like many incompatible couples forced together by circumstance rather than choice, they have worked out an amicable coexistence and have even become rather dependent on each other in their old age.

The smell-brain does not get much publicity, which is a shame as it contains some of the most varied epithets in the entire cartography of the brain. Although small in humans, its terrain is diverse—a sign of the importance of smell in our more odor-obsessed ancestors. In many animals, the smell-brain protrudes proudly from the front of the brain, but in humans it is more reclusive, nestling beneath the looming mass of the cerebral cortex. You may remember that in Chapter 12 we saw how the nerves from the nose connect to the left and right olfactory bulbs, which then carry out an elegant feat of computation to provide an instant chemical analysis of the smelly things whiffed into the nose. Beyond the bulbs, smell information is conveyed backward, ramifying into many diverse areas of the brain. And as you may remember, it is unique among the senses in claiming direct access to our conscious mind, being able to enter the cortex directly without any prior perusal by the drily analytical thalamus.

Each of the bulbs sends back two bundles of nerve fibers carrying smell information—an inner and an outer stria, or "groove," visible on the underside of the brain. The two stria diverge from each other, creating the apex of a triangular region called the olfactory trigone. Between the stria lies the vaguely named anterior perforated substance, a region pierced by many small blood vessels. Embedded within this tissue is an archipelago of scattered grey matter, the islands of Calleja—cup-shaped atolls of neurons, usually one large one and several smaller on each side. The islands are probably tiny, isolated gobbets of internalized cortex, as they seem to share its cellular structure. Behind the anterior perforated substance is a region where the smell fibers start to fan out to their various destinations in the brain, the radiations of Zuckerkandl, named after an Austrian neuroanatomist working at a time when neuroanatomists were clearly in vogue—he counted Rodin, Wagner, and Klimt among his friends.

From the smell-brain, the inner and outer stria course to many destinations. Some fibers may switch between right and left across an ancient bridge called the anterior commissure. It has been claimed that this allows some animals to compare the smells arriving in each nostril to determine from which direction a smell is coming. Obviously, the differences between the smells in the two nostrils is greater if they are farther apart—a possible reason for the shape of a hammerhead shark. Some fibers pass to areas of the forebrain nestling between the two hemispheres—the parolfactory area of Broca (Broca's brain is now pickled in glass jar) and the septum pellucidum, the "clear boundary hedge." The cleft between the two sides of the brain in this region is, for some reason, called "the cave." From the environs of the pellucid cave, smell information is then relayed to the hypothalamus, that controller of our most visceral functions. In particular, it travels to the preoptic area, which we have already seen is a different size in men and women and controls sexual behavior. Already smell has penetrated to our inner lusts.

From the stria, other smell information is passed to a small, pear-shaped region of cortex on the underside of the brain that I mentioned earlier called the piriform lobe. The piriform lobe is probably mammals' main vestige of the side pallium, and thus is our diminutive equivalent of the large reptilian dorsal ventricular ridge and avian wulst. From the piriform, smell impulses are once more dispatched far and wide. Some go to the nearby amygdala, a region involved with emotion and fright to which we shall return—the smell of fear, perhaps? Others travel to the habenula, which you may recall was that hump on top of the interbrain involved in the "brain reward system" and possibly addiction—the smell of chocolate, maybe?

Finally, there is an extremely circuitous additional route by which smell can reach the cortex. After performing a loop-the-loop through some of the forebrain's most tangled terrain, these fibers end up in the mammillary bodies, a pair of bumps at the back of the hypothalamus. I have always thought that this name gives a sad little insight into some frustrated neuroanatomist's lonely life, as it means "breast-like," and I have to say that the similarity is only slight. The breasts themselves are involved in controlling aspects of our behavior (gentlemen readers may wish to cogitate on that), but they also send fibers up to the cortex via the superbly magniloquent "mammilothalamic tract of Félix Vicq d'Azyr," named after the eminent eighteenth-century French comparative anatomist and veterinarian who discovered it. Personal physician to Marie Antoinette, he also discovered the "sky-blue place," the locus coeruleus, and was the first person to map the convolutions of the cortex in detail, although not in his most famous patient, it is hoped.

So the smell-brain has sent its tendrils throughout our forebrain. In some vertebrates those tendrils are so important that they make up a large proportion of that forebrain—the tendrils are relatively thicker and the space between them is smaller. But there is always space between those tendrils in which the forebrain can carry out its other role to act as the "higher center" of the brain. In many vertebrates, there has been only a limited tendency to concentrate sensory analysis and motor control in the forebrain, but in mammals, and especially humans, that process of cephalization has been dramatic. As a result, our nonsmell forebrain now completely overshadows our smell-brain. And the cerebral hemispheres of humans are among the largest in existence.

Yet at first sight the cortex is surprisingly homogeneous. As we have seen, it is a thin layer of grey matter coating the outside of the cerebrum, and Cajal demonstrated that the neurons of the bulk of the mammalian cortex are stratified into six layers. In fact, this six-layered cortex is a distinctive feature of mammals. When he explored what he called "the apparent disorder of the cerebral jungle," he found that, unlike the long, spindly nerve cells in the rest of the brain, many cortical neurons are quite short, stubby things. They often have short dendrites and short axons, and presumably carry out most of their interactions with their nearby neighbors. They are often described as pyramidal because of the shapes of their cell bodies—yet another occurrence of the "pyramid" theme in neuroanatomy. Cajal speculated that our highest functions are played out in this densely layered sheet of cortical neurons, and indeed we now suspect that these neurons have electrical properties that make them especially good at collating multiple simultaneous inputs. Unlike the rest of the brain, in which impulses are purposefully channeled over long distances to their predetermined destinations, the cortex may be more like a fizzing, rippling, unpredictable sea of electrical activity.

A fizzing electrical sea is, however, heresy to anatomists. It raises the awful possibility that higher brain functions cannot be localized to particular regions. What if the activity of the cortex is carried out in large, diffuse circuits that all overlap with each other? Neuroanatomists would be out of a job. The vague claim that perception, emotion, and intellect simply take place "up there" in the cortex was not precise enough for the mapmakers of the mind, and even before Penfield, neuroscientists had been charting the convoluted terrain of the cortex ever since Félix Vicq d'Azyr. There have been several systems invented to subdivide the cortex into some sort of ordered arrangement. One used most often is a numerical system invented by the German neuroanatomist Korbinian Brodmann, in which cortical regions were allocated a number: "area 1," "area 2," and so on. This subdivision is still used today because it provides a universally accepted system of cortical territories to which scientists can refer. It is, however, completely lacking in the poetry that suffuses the nomenclature of the rest of the brain, and it is not even logically ordered. Rather than basing the system on the location of presumed function of areas of the cortex, Brodmann simply numbered his regions in the arbitrary order in which he studied them. Yet a glimmer of intrigue persists. UFO conspiracy theorists will be pleased to hear that "Area 51" is not only a legendary secretive military area in New Mexico, it is also the periamygdaloid cortex, next to the fear- (and paranoia?) inducing amygdala.

Another approach was to name cortical regions according to the convolutions on the surface of the brain (see Figure 4.7). First noticed by the ancient Egyptians all those millennia ago and compared to the corrugated scum that forms on molten copper, the ridges and valleys of the hemispheres have fascinated us ever since. It has long been noted that convolutions are present on the brains of animals that appear to be more intelligent than their smooth-brained relatives, and the presence of convolutions has often been equated with sentience. However, there are other reasons why the cortex may form furrows, and the irregular surface of our own cortex is probably just a way of increasing its surface area. This corrugation seems to have been especially effective in humans, as only one-third of your cerebral cortex is on the outside of the brain, whereas two-thirds is hiding deep in the furrows.

The bulging, sinuous ridges of the cortex are called gyri, or "coils," and the valleys are sulci, or "furrows." Sometimes the sulci are called fissures, especially if they are large, although there is no clear distinction between a sulcus and a fissure. Some are shallow, whereas some are so deep that the infolding of the cortex impinges on the fluid-filled lateral ventricle at the core of the hemisphere. One such characteristic intrusion occurs at the very back of the ventricles where a ridge called the calcar avis, or "bird's spur," presumably referring to the fighting spurs of cockerels, is visible. The calcar is the inward sign of a deep cleft called the calcarine fissure, around which Penfield found that he could induce visions of flashing or moving lights in his subjects. We now know that the cortex around this fissure is where visual information is first represented—the primary visual cortex. This region also varies among individuals—the morphology of the nearby lunate fissure of Altenspalte was once used rather dismissively as a characteristic of "lower" human races. Despite these minor variations, the pattern of gyri and sulci is actually remarkably consistent among different people, suggesting that their formation is an ordered process, rather than the result of an expanding cortical sheet randomly crumpling as its growth is hemmed in by the skull.

The most commonly used geography of the cortex is that of the lobes (Figure 17.2a). Yet although this subdivision is the best known, chopping the cortex into frontal, parietal, occipital, and temporal lobes is really rather arbitrary. The cortical sheet of each hemisphere is not really divided into four sections, but is instead a single, continuous whole. To shore up this artifice, we selected some especially deep fissures to act as boundaries for the lobes, but that sleight of hand becomes evident when we study other mammals, since none

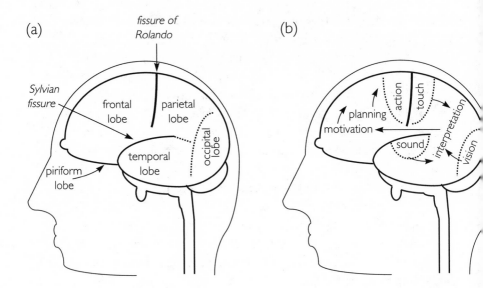

(a)

fissure of Rolando

Sylvian fissure

frontal lobe

parietal lobe

occipital lobe

temporal lobe

piriform lobe

(b)

planning

action

touch

motivation

interpretation

sound

vision

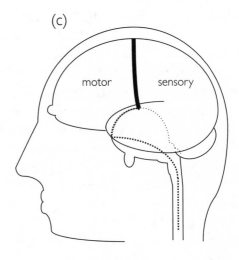

(c)

motor

sensory

of them share our pattern of gyri and sulci. Yet the four-lobe concept has stood the test of time, partly because the lobes have fortuitously turned out to have surprisingly distinctive functions. These roles had been suspected for some time, largely because of evidence from people who suffered injuries or disease of the different lobes, but Penfield's experiments were the start of a detailed investigation of the lobes that is still continuing today.

Immediately after Penfield, it was beginning to look as if a blissfully simple scheme might emerge. The primary sensory regions, where sensation first reaches the cortex, had been detected—touch at the front of the parietal, vision at the back of the occipital, and hearing in the temporal lobe near the deep Sylvian fissure (Figure 17.2b). From these regions, the different sensory strands are transmitted to new areas where they are processed and analyzed further, first in isolation from each other, and then all together to yield a combined, contextualized view of the world. From here, this view of the world is sent forward to the frontal lobes, where it is used to generate motivations, plans, and eventually actions by activation of the primary motor cortex, which lies at the very back of the frontal cortex. So information comes in at the back, is analyzed, sent to the front, and converted into actions. If only it were that simple.

I would like to stick with this simplistic scheme a little longer

---

Figure 17.2. (a) The lobes of the human cerebral cortex. Some lobes are separated by obvious fissures or clefts—such as the fissure of Rolando that separates frontal from parietal and the Sylvian fissure, separating temporal from frontal. The boundaries between others are more arbitrary. (b) A simplified view of how the cerebral cortex works. The primary sensory regions, where sensation first reaches the cortex, are in the back half of the hemispheres. From these regions, the different sensory strands are transmitted to new areas where they are processed and analyzed further, first in isolation from each other, and then all together to yield a combined, contextualized view of the world. From here, this view of the world is sent forward to the frontal lobes, where it is used to generate motivations, plans, and eventually actions by activation of the primary motor cortex. (c) An even more simplified view of the brain in which the arrangement of cells in the spinal cord—sensory at the back and motor at the front—is continued all the way up to the cortex.

before we pick it apart, because it fits rather neatly with a theme that has traveled throughout this book. This idea of sensation and interpretation at the back and planning and action at the front has a reassuring concordance with the way we looked at the spinal cord and brain stem. You may recall that down in the spinal cord, the sensory cells were clustered at the back and the motor cells were clustered at the front. I also argued that this arrangement continued, with some distortion, all the way up through the brain stem. Even through the peeling open of the hindbrain and the ninety-degree kink in the midbrain, the system worked in a rough and ready way. Now we are up in the cortex, and if we allow another ninety-degree kink, it could be argued that the scheme still works up here (Figure 17.2c). The possibility that we could have essentially the same format throughout the entire nervous system is so philosophically attractive that it tempts us to believe it is true. And despite its problems, I still believe that it is a good way to think about the cortex. Some ideas are so clear and simple that they can help us even if we are worried that they may not be entirely correct.

The first problem with this scheme has to do with the weird way the endbrains develop in the embryo. The brain stem and spinal cord form with a back part and a front part, clearly distinguishable from when the nervous system first curls up into a tube. And it is these two regions that will contain the sensory and motor cells, respectively. However, the endbrains simply do not fit into this scheme. They bulge out from either side of the developing brain after the back-front differentiation has already been established, and there is no evidence that they are subdivided in a similar way. As a result it has been suggested that the endbrains are entirely derived from the back, "sensory" region of the neural tube and that this is why no such subdivision is ever present. To be honest, the endbrains may form in a way so unlike the rest of the brain that it is misleading even to attempt to fit them into the same scheme.

Another problem has to do with the evolution of the mammalian brain. If the distinction between sensory at the back and motor at the front is an intrinsic feature of how the brain forms, then you might expect it to occur in all mammals. Certainly it holds true in humans, primates, and most of the domestic and laboratory animals, but even among these the neat distinctions between regions and functions can fall apart. Yet if we look at the other great group of mammals, the marsupials, we find an alarming merging of sensory and motor function in the cortex. For example, in the marsupial brain the primary sensory and motor cortices—the two areas that in us are most discrete and consistent in their arrangement—are mixed and merged together. We could counter this problem by saying that marsupials have distorted an older, more ordered system, were it not for the fact that one group of "placental" mammals have a similarly mixed-up arrangement. These are the edentates—the toothless sloths, anteaters, and armadillos—a cluster of species that are quite remotely related to the rest of the placental mammals like us. The fact that both these and our more distant cousins, the marsupials, both share this vague cortical arrangement suggests that this is the original ancestral system which the rest of the mammals subsequently altered. So, the neat clustering of sensory activity in one region and motor activity in another is a novel development shared by some placental mammals: an arrangement of pragmatic convenience rather than deep underlying principle.

And when we look more closely at the activities going on inside the modern human brain, we find even more niggling exceptions to the rules. The functions of the supposedly sensory parietal lobe are especially confused. The areas of this lobe outside the primary sensory areas certainly have some complex jobs to do. As Penfield suggested with his "psychical" effects, the back part of the parietal lobe is probably involved with high-level perception and interpretation, but it now seems that those functions may blur into more active, motorlike activities. For example, this area is probably involved in the control of attention, the focusing of the mind on a selected feature of the world outside, as well as intention, the direction of motor activity based on what is being concentrated on. As you can see, we have moved up to some fairly high-level stuff.

The temporal lobe also presents us with such apparently complex functions that it is increasingly difficult to cram them into the sensory category. Not only does this region seem to confer familiarity and unfamiliarity—hence Penfield's patients and their *déjà vu* and *jamais vu*—but diseases of the temporal lobe can cause dramatic and unsettling effects. The deeper areas of the temporal lobe are very close to some unusual areas at the core of the hemispheres, and this may explain their penchant for strange phenomena. One of these regions is the slab of cortex buried inside the deepest fissure of all, the Sylvian fissure. Occasionally elevated as a lobe in its own right, this invisible expanse of deep, hidden cortex is the insula or island of Reil—the area Nick was looking at in my brain in his study of disgust and depersonalization. Maybe these are the realms where we understand who we are. It has even been suggested that undiagnosed temporal-lobe epilepsy may explain the religious experiences of the saints and prophets of history. Perhaps all we need is someone with temporal-lobe epilepsy as well as that strange form of synesthesia in which they see haloes around people's heads, and we will have a full-blown messiah on our hands. Forget the "God gene": the parietal might be the "God lobe."

The frontal lobes are, if anything, even more of an enigma. They carry with them the paradox that although they are the biggest lobes of all, large chunks of them can be removed with apparently minimal effects. You may remember that Penfield removed most of one of his sister's frontal lobes—she was able to function perfectly well, as long as major planning decisions were made for her. And there are several examples of people who accidentally sliced or blasted off most of their frontal lobes and suffered few lasting after-effects as a result. Much of the frontal lobes seems to have a high-level role in planning ahead, and this is fed back to the motor cortex at the back of these lobes when something needs to be done. All of this planning seems to be linked to anxiety and strong emotional responses, and this was the basis of the frontal lobotomies once fashionable for sufferers of various psychiatric disorders, as well as some less-mentioned members of the Kennedy dynasty. Slicing these lobes away from the rest of the brain does not make people blind, deaf, paralyzed, or stupid. If anything it damps down emotional responses and makes people a little irresponsible. They focus on the here and now and have little drive to look beyond the obvious. They may behave inappropriately in social situations and simply not consider what others think. Thus, it is tempting to say that the frontal lobes are for planning ahead—something at which humans are especially good—but that would be far too simple. For example, some regions of the frontal lobe seem to be explicitly involved in the interpretation of sensory information, and in particular the experience of pleasure. Thus, as a region where pleasure and planning ahead lie side by side, the frontal lobes are also being actively investigated for their role in addiction.

The greatest inconsistency in my scheme of the brain lies at the back of the frontal lobes in the primary motor cortex. I have implied that instructions are sent back to this strip of neurons that then relays the deliberations of the frontal lobes down to the body as decisive actions. Unfortunately, the motor control of the body is not as simple as that. First of all, in many nonprimates the motor cortex is far less important—in domestic animals, most motor commands seem to arise from the red nuclei in the brain stem. Second, controlling movements is an extremely complicated business. Some movements are planned in advance and then enacted without pause for thought. Other movements are slower, guided, and constantly corrected if they go astray. And whatever the type of movement, a tremendous amount of sensory information is required. Are my eyes pointing at the pen? Where are my eyes pointing? Is the pen too far away to pick up? Is it moving? How is it lying? Where is my arm? Am I moving? Is my hand open? There is a bewildering amount of computation involved in completing even the simplest task, something that engineers discover when they try to design robots.

Little surprise, then, that the motor cortex has some extra computing power to help it. We still do not know exactly how they contribute to making movements, but the two main additional motor areas comprehensively mess up our neat scheme of brain arrangement. The first of these is the cerebellum, the globular "little brain" that grew out of the top of the hindbrain (see Figures 4.6d, 4.7, and 9.3). In nonmammalian vertebrates the cerebellum is already a processing center where sensory information is used to guide movements, but in cerebral mammals it spends most of its time interacting with the cortex in the control of movement. Like the cerebrum, the smaller cerebellum has most of its neurons on its external surface in the cerebellar cortex, yet the cerebellar cortical neurons are even more regularly arranged and densely packed than those in the cerebrum. Also, the surface of the cerebellum is even more convoluted than the cerebrum and as a result appears finely ridged, as you may notice in my MRI scan (Figure 4.7). These fine ridges are bunched into various larger agglomerations, fancifully called the vermis, flocculus, nodulus, lingula, culmen, declive, tuber, and uvula—the "worm," "smidgen," "little knot," "little tongue," "summit," "slope," "hump," and "little grape." There is even yet another "pyramid." The contribution of the cerebellum to posture and movement becomes clear when it is damaged. A few years ago, my nephew James showed cerebellar symptoms as part of a post–chicken pox encephalitis. His walking soon became uncoordinated, he lost his balance and fell over, his eyes would drift and flick, and he trembled disconcertingly whenever he tried to make a planned movement. All this was terrifying to his parents even though they were advised, correctly as it turns out, that the effects would be temporary. To three-year-old James, of course, it was all extremely hilarious.

The other region of the brain that helps to control motor movement is the basal nuclei (sometimes called the basal ganglia). These are the chunks of basal pallium that invaded the center of the mammalian brain (Figures 17.1 and 17.3). They also appear to be involved in modulating and controlling movements, and we have already seen how Parkinson disease results from a failure of the substantia nigra to control the activity of these structures. There are several basal nuclei on each side, and we are gradually working out how each contributes to making smooth, even movements. The innermost is also the largest and is called the caudate, or "tailed" nucleus. It consists of a globular mass of grey matter with a long, spindly tail. Outside this lies the lentiform or "lentil-shaped" nucleus, although "brazil nut–shaped" might have been more appropriate because its shape resembles a brazil nut laid on end. Closer anatomical examination reveals that the lentiform is divided into subsections, which also seem to have discrete functions in controlling movement. These are the putamen, or "husk," and the internal and external portions of the globus pallidus, or "pale globe," also sometimes called the rather papal-sounding pallidus I and pallidus II. You may recall that one of the more aggressive treatments of Parkinson disease is a partial assassination of one of these two. Beyond the lentiform complex lies the claustrum, or "barrier," although this may be an isolated slab that has recently split from the overlying cortex of the island of Reil rather than a true basal nucleus—a sand spit of grey matter cut off by a rising tide of white.

Partly because diseases of the motor system are so common, many neuroscientists spend their time trying to work out the relative contributions of the basal nuclei, cerebellum, and cortex to movement. At present, motor systems are inherently difficult to study and their answers are complex and uncertain, but one thing is clear. The regions involved are scattered throughout our heads in a way that rather contradicts our neat little theory of the organization of the brain.

Our scheme may be looking a little frayed around the edges, but I hope I can convince you that it is still a good way for us to start fitting something as complex as the mind into the wiggly, unprepossessing hemispheres. Despite all the caveats, it is still fair to say that sensory information comes in at the back of the cortex, is processed to ever higher levels of perception and interpretation, and these are then used to plan and motivate our actions in the front of the cortex (presumably after some input from our memories and emotions, which we have not yet considered). The fact remains that a large fraction of your cortex is not involved in low-level analysis of sensory information nor the mundane job of activating your muscles. Most of it is doing the clever stuff in between—linking perception to action. We call this high-level linking "association," and it has become a concept that drives much of our modern thinking about the brain. Extensive areas of the cortex are now called "association cortex"—the large region at the junction of the occipital, parietal, and temporal lobes at the back of the brain where Penfield found his "psychical" effects, and another large region at the front of those huge frontal lobes. Association has been important because it has allowed us to separate the simple and complex functions of the brain. It lets us distinguish our higher functions from everything else, both in terms of specific anatomical regions and as a phenomenon that is engineered within those regions. Association takes us a defined, tangible step closer to that slippery thing called consciousness.

Many scientists now agree that the higher associative functions of our brain can be localized to particular regions in the same way that can be done with the twitch of a muscle or a tingle on the skin. This

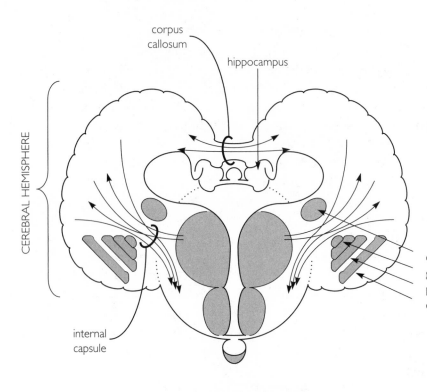

corpus
callosum

hippocampus

CEREBRAL HEMISPHERE

caudate
globus pallidus
putamen
claustrum

internal
capsule

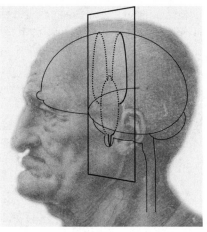

does not mean that they have to be located in the same places in different people, and certainly not in different species. It also does not mean that those scientists can agree on where all those locations are—after all, that would put them out of a job. However, there is a general feeling of a sense of place, that quite esoteric and abstract functions of mind are played out in a particular gyrus or sulcus. In addition, most do not feel that there is any insurmountable boundary between association and other more esoteric concepts like feelings, understanding, and consciousness—all should be amenable to the same methods of study. Of course, this is music to the ears of anatomists, who had always feared that these activities might be coordinated by large, overlapping circuits of cells spread over large stretches of cortex. Forget the fizzing electrical sea. A place for everything and everything in its place—that is what we like.

Although it may seem like jumping the gun a little, there has always been a suspicion that association actually is the same thing as consciousness. As we will see, consciousness is a difficult concept to nail down. We have no adequate definition for it, and we have no objective way of measuring it. Yet some neuroscientists still propose that it may be the result of high-level association—what is consciousness, if not the way that we use all our brainy faculties to peruse and respond to the world around us? I do not want to come down on either side of the argument yet, but there is something to be said for all this. It gives consciousness a place and a definition, although it could be said to lower its philosophical status somewhat. Then again, we are happy to think of the nervous system of 'lower' animals as being

---

Figure 17.3. A cross section of the forebrain with the structures of the cerebral hemispheres emphasized. Below is a profile (by Leonardo da Vinci) overlain by an outline of the brain. The rectangle is the approximate level of the section. The central circular region is the interbrain out of which the cerebral hemispheres have grown on either side. At the center of the interbrain is the tall, narrow third ventricle, which communicates with the lateral ventricles of each hemisphere. Deep in each hemisphere lie the basal nuclei, and between these runs the internal capsule, carrying long axons to and from the cerebral cortex. In mammals, there is an additional link between the two hemispheres at the top, the corpus callosum.

there to generate responses to the environment, so why should not our brain be doing just that, albeit in a more complex way?

And much of the experimental evidence is adding up to support this view. Regions of cortex are gradually being ascribed defined functions, some verging on what we think of as consciousness. Soon there may be no brain left where a more spiritually defined "soul" or "mind" can hide. Association cortex is itself a layered concept. Some regions are obviously lower level than others, storing motor "programs" to carry out complex movements or making quick, learned reactions to emergency situations. Others seem slightly higher, such as mediating anticipation of events in the world outside. Elements of memory are also up here, but memory is a subjective and selective thing—long-term memories may be held in the association cortex at the back of the brain, but it may be the frontal association cortex that coordinates the retrieval of memories from that cerebral filing cabinet.

Most strikingly of all, diseases that affect the association cortex can strike at the very core of the self. We have already mentioned the strange world of depersonalization disorder, and imaging studies have shown that it is linked to changes in activity in the island of Reil as well as the association cortex at the back of the brain. Also, some of the most disturbing changes caused by Alzheimer's disease occur when blood flow, activity, and neuronal connections in the association cortex are compromised. In addition, whether or not epileptics lose consciousness during their seizures seems to relate to whether or not the electrical storm affects their association cortex. More restricted injuries to the association cortex tell a similar story. They frequently do not cause symptoms that can be detected by neurological tests, so patients are often declared to be completely recovered. However, they often suffer work-related, social, or marital problems in the aftermath of their injury, possibly as a result of more subtle alterations to how they interact with the world. And in an example that shows how surprisingly important perception is in the normal functioning of the self, some injuries can cause people to completely ignore entire aspects of the world, such as one side of their body, even though they can sense them perfectly well.

So there is considerable evidence that association is a phenomenon that reaches at least as far as generating a sense of the world, a sense of self, and a basis for social interaction. These relatively abstract concepts seem to have a real physical existence up in the cortex, just waiting for us to find them. And in a confusingly self-referential way, we humans probably needed to evolve sufficient powers of perception, interpretation, and planning to become intelligent enough to discover them. The story of human evolution seems to have been the story of formation of more and more association cortex. Other mammals do have it, but we have more of this arcane, self-analytical stuff. This should not surprise us—we never thought that humans could sense or move any better than other animals. We simply wondered if there was more cogitation going on between sensation and action. Maybe Galen's bridge is simply higher in us.

The idea that the cortex is an array of sensory regions linked up to motor regions by a series of association regions has changed the way we think about its structure. It was always tempting to think that it was the grey matter, the neuron cells themselves, that was important. This even leaked into common parlance—we speak of intelligent people as having more "grey matter." Yet the idea of association and the anatomical fact that all the grey matter in the cortex looks rather similar have made us change our view. We now realize that it is probably the pattern of connections among the neurons that decides how the brain works. It is the branching and cabling of the white matter, the axons, that controls how different parts of the cortex interact with each other. Underlying the cortical grey is a larger region of white matter, a tangle of fibers connecting neurons over distances long and short, like an exaggerated version of the jumbled wires that connected the sockets on an old-fashioned telephone switchboard. "I'll put you through, Mr. Parietal."

These billions of axonal connections can be divided up into three main groups: within cortex, cortex to brain stem, and cortex to cortex. The first of these, the within-cortex fibers, are also called the association fibers, which shows how that concept has become embedded in our thinking. They leave one patch of cortical grey by plunging into the underlying white, and once there they travel to their intended destination in a region of cortical grey elsewhere. Some association fibers are very short, connecting two neurons within the same gyrus or in adjacent gyri. Others can carry information from one end of the brain to another. These longer fibers are often arranged into large cords of white called fasciculi, or "bundles," which arc around inside the hemispheres. Although deeper and less accessible to study than the overlying grey, we assume that it is the arrangement of these association fibers that in large part defines how the cortex works.

The function of the next type of fibers, the cortex to brain stem fibers, is probably self-evident. Apart from smell, the cortex receives all its information via the brain stem, and it issues its motor edicts via the brain stem to the body. For example, the sensory lemnisci must get to the cortex and the motor pyramids must get out. Thus, there are very large bundles of fibers traveling up to the cortex from the stem—the left and right internal capsules. The capsules are so large that they are clearly evident in a brain sliced from side to side (see Figure 17.3) as they squeeze between the lentiform nuclei on one side and the thalami and caudate nuclei on the other. As you might imagine, the internal capsule is an especially dangerous area for disease or injury. A small stroke in this area can wipe out much of the connection between a hemisphere and the body. And as they ascend up to the cortex, the tight bundles of the internal capsules spread into the descriptively named coronae radiatae, the "radiating crowns." The coronae fan out in all directions, dispersing the axons in the capsules to all their diverse cortical destinations. Thus the cortex is not unlike a veil cast over a crown of axonal thorns.

The third type of white-matter connection is the cortex-to-cortex axons, also called the commissural fibers from the Latin for "uniting." The need for special left-right connections results from the way the hemispheres develop in the embryo, leaving the left and right sides largely separate from each other—the endbrains bulge out of opposite sides of an interbrain that is itself largely bisected by the deep cleft of the third ventricle. It is intriguing that the largest part of our brain looks as if its two halves were almost designed to be kept apart, and neurobiologists have long wondered why this is so. The commissural fibers are the exception to this rule, as they are the axons that actually link the two sides together. In nonmammalian vertebrates, there are three main commissures—the posterior one that we mentioned fleetingly, giving its name to that strange fiber-exuding subcommissural organ, the anterior one that we name-checked as it transferred smell information from side to side, and the commissure of the fornix, which we will worry about later. And in most verte-brates, these scanty threads seem to be sufficient to join left and right.

However, as mammals evolved their relatively larger forebrains, these commissures were not sufficient to bridge the ever-expanding hemispheres. A new link was created between left and right (Figure 17.1), sprouting from the top of the interbrain and then growing backward. Egg-laying and marsupial mammals do not have it, but placental mammals like us now have a dense, tough strap of white matter linking our hemispheres: the "hard-skinned body," or corpus callosum (Figure 17.3). The corpus callosum is huge compared to the other commissures—its dense, pale mass is one of the most obvious features of my own MRI, for example (see Figure 4.7). This new bridge between the halves of our brain has become central to the way we are organized, but it carries its own contradictions. First of all, why did we form it, rather than simply fuse the two hemispheres as some other vertebrates have done? Yet the corpus callosum does not seem to have evolved as a first step toward fusion of the hemispheres, because now that we have it there is no sign of further fusion in living mammals. Communicating bridges are acceptable, but it seems to be best for the hemispheres to remain distinct. Why? A second strange feature of the callosum is that such an important structure is so vari-able in size among individuals. This variability does not seem to be related to any measurable characteristic or ability, except for the un-explained fact that the back part of the callosum, the splenium or "eye-patch," is larger in women. Also rather unnerving is the fact that cutting the corpus callosum—as is occasionally done to prevent sei-zure activity spreading from one hemisphere to the other—has little obvious effect on the functioning of the brain. These so-called split-brain patients do not behave very differently, despite the fact that the main cable between their two hemispheres is severed—at least not until we start doing strange experiments on them, as we will see.

These commissural complexities lead us on to the one last aspect of cortical structure we need to address before we can plunge inward to a unified view of how the brain is put together—the nature of the right and left sides of the cortex. In fact, there are two intertwined stories we must unravel here. The first one we have already encountered—the fact that the forebrain seems to be wired the wrong way round. The right hemisphere deals with information from the left side of the world and moves the left side of the body, and vice versa. In contrast, the lower parts of the brain are more sensibly arranged, so every axon ascending to or descending from the cortex must switch sides at some point, hugely adding to the complexity of the cabling in the brain stem. We are fascinated by the inversion of the forebrain and the way the world is represented within us in mirror-image form, but we can see no good reason why such a strange arrangement should be beneficial. Scientists do not like to accept this sort of thing at face value, as it contravenes their ideas of natural parsimony. We like to think that things in nature are as simple as they can possibly be, not because they were designed by some omniscient creator, but because simple systems evolve more readily and are more resilient. Yet with the inversion of the forebrain, we have a biological system that is clearly more tangled than it needs to be. To solve this conundrum, we must look back to the evolution of the vertebrate nervous system.

All vertebrates flinch away from unpleasant things. If you were to poke a goldfish on its right flank, you would find that it bends toward the left, away from the noxious finger. Thus, there must be a reflex link between sensory receptors on its right side and the contraction of the muscles on its left side, as it is these muscles that will cause the evasive bending of its body (Figure 17.4a). To make it easier to coordinate the response, the two neurons involved—the sensory one and the motor one—are near each other. They could both be in either side of the brain, but it seems that by evolutionary chance they were both placed on the left-hand side, farther from the poke, but nearer to the muscle. However, we no longer live the fishy life, and we walk about on dry land using legs, even though our brains evolved from those of fish. When a four-legged land vertebrate is prodded on the right side, it is just as eager to escape its tormentor, but it does so by straightening its right limb—the limb on the side of the poking (Figure 17.4b). It seems that the reflex system it uses to cause this straightening evolved from the old fishy system, as both sensory and motor neurons are still on the side of the brain opposite the stimulus. Once they were there, it was apparently easier to send the motor fiber back

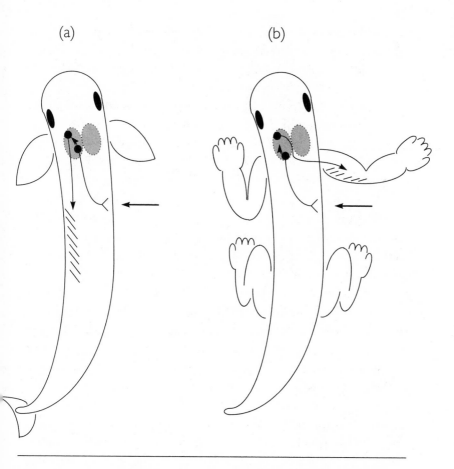

(a)   (b)

Figure 17.4. A theory for why the mammalian forebrain is wired up backward. (a) If a fish is touched on its right flank, it bends toward the left, away from the noxious stimulus. Thus, there is a reflex link between sensory receptors on its right side and the contraction of the muscles on its left side. The two neurons involved—sensory and motor—are near each other, both placed by evolutionary chance on the left-hand side. (b) When a four-legged land vertebrate is touched on the right side, it straightens its right limb. Because the reflex system it uses to cause this straightening evolved from the old system in fish, both sensory and motor neurons are on the opposite side of the brain from the stimulus.

across to the right than to move all the neurons from one side of the brain to the other. Thus we have a right-sided stimulus leading to a right-sided muscle contraction, but both neurons are on the left side of the brain.

Is this truly the reason why the forebrain is backward—the adaptation of a response in fishes for use in leggy land vertebrates? It certainly ends up with sensory and motor fibers having to cross as they travel to and from the forebrain. Also, apart from the random positioning of the two neurons in the fish brain, the story does seem to make sense. According to this theory, the wrong-sidedness of the complex human cortex is a relic of a simple, protective reflex in our distant fishy ancestors. Once wired into the forebrain, this inversion could not easily be reversed, and instead it was built upon and enhanced. Once again we see evolutionary historical accident rather than intelligent design in the formation of our brain.

Superimposed on this ancient inversion of the two cerebral hemispheres is the discovery that they differ from each other in both structure and function. The first evidence of this asymmetry came from the simple observation that the hemispheres are lop-sided—the front of the right hemisphere is larger than the left, and the back of the left hemisphere is larger than the right. The right frontal lobe protrudes farther forward and even bulges a little more out to the side—the so-called frontal petalia ("petalon" is Greek for "leaf," and its plural is "petala," so this is a rare example of linguistic incompetence by anatomists—there is no word "petalia" in any classical language). Conversely, the left occipital lobe bulges out the back and slightly to the side—the occipital petalia. This rather dramatic wonkiness of the cortex has even been suggested to be the result of a twisting of both hemispheres about a vertical axis through the center of the head—as if some mischievous god reached down and sloshed your cerebrum around a few degrees anticlockwise. This alarming idea even has its own alarming name—the Yakovlevian torque ("torquere" is Latin for "twist").

The Yakovlevian torque does have some other evidence to support it. For example, the left occipital lobe at the back of the brain can bulge across, actually displacing the right occipital lobe somewhat. Because of this, the curtain of meninges that separates the two sides, the falx cerebri, may be deviated slightly to one side. Some of the most dramatic asymmetries are present around the Sylvian fissure, and these too could be argued to fit in with the idea that the whole brain has been rotated within the skull. However, when the two hemispheres are compared more generally, the twist does not seem to be so generalized. Large regions do not show consistent differences between left and right, and certainly no simple rotation within the skull. The torque remains a challenging idea, but one that is far from accepted.

One region where you might expect some dramatic asymmetry is the back of the frontal lobes, where the primary motor cortex lies. This is the start of the pyramidal tracts, which control fine, planned, manipulative movements. The majority of the human population is right-handed or left-handed, and so one might expect the cortex that controls the preferred hand to be better developed. Yet surprisingly the differences between the left and right motor areas are generally considered to be extremely small, if they exist at all. But to add to the confusion, some data suggest that the Yakovlevian torque (if that is what it truly is) is more pronounced in right-handers.

And why is the region around the Sylvian fissure so asymmetrical? The fissure itself is usually longer on the left, and one nearby region, the planum temporale or "temporal plane," may be up to ten times larger on that side. Excitingly, these are areas we think are important in understanding and producing language. As long ago as the nineteenth century two neurologists, Paul Broca and Carl Wernicke, reported that diseases of the left side of the cortex are much more likely to cause problems with language. Today, Wernicke's and Broca's areas are known to be two of the main language-processing centers in the brain. We will return to language later, but to simplify matters considerably, Wernicke's region is involved in the interpretation of speech and lies conveniently close to the primary hearing cortex, whereas Broca's region coordinates speaking and is next to the primary motor cortex for the mouth (Figure 17.5). But what is remarkable is that in most people these linguistic functions are clustered exclusively in the left hemisphere. These areas are on the left in 97 percent of right-handed people and 70 percent of left-handed people—a difference that in itself raises interesting questions. Yet the overriding question

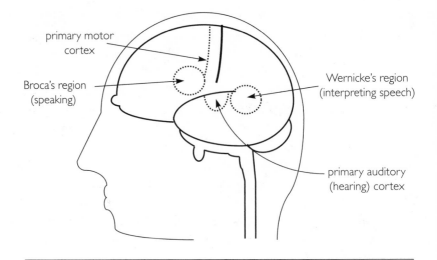

Figure 17.5. Some cortical regions involved in language. Wernicke's region is involved in the interpretation of speech and lies conveniently close to the primary hearing cortex. Broca's region coordinates speaking and is next to the primary motor cortex for the mouth region.

is why, among all the functions of the human brain, language should be so exceptionally asymmetrical. I have to admit that we do not really know, but we do have some theories. Maybe left and right language areas would conflict with each other, or maybe it is simply more efficient for all the processing to be concentrated in one small area on one side of the cortex.

Studies with split-brain patients, who have undergone sectioning of the corpus callosum to separate the two hemispheres, have shown that the right and left brains differ in more ways than just the linguistic. Experiments can be arranged so that objects are shown to one hemisphere but not the other, and one hemisphere can be asked to carry out tasks via the hand it controls on the opposite side. As you might expect, the left hemisphere seems to be the pushy, mouthy one. It listens to you, talks to you, reads things, writes things, does arithmetic. It is articulate and analytical, but not profound. In contrast, the right hemisphere can initially appear submissive and uncommunicative. However, it is very aware of the world around it, it can draw but not write, it has a good sense of perspective about things, and it can make exciting intellectual leaps. Most disturbingly of all, if one hemisphere discovers that the other hemisphere has been trying to carry out a task for which it knows it is better equipped, it can get quite affronted and use its hand to push the hand controlled by the other hemisphere out of the way.

In some ways these patients can be made to behave as if they have two separate brains in their head, and in a very real way they do. Of course you do too, but they are connected a little better. Just as the genius of the Beatles resulted from the conflict between Lennon and McCartney, is the wonderful human intellect a result of the arguments between the two dissimilar inhabitants of our skull? Is this why the two sides of the cortex are forever kept apart—to maintain their individuality and even to make them argue?

For many years we thought that right-left asymmetry was a distinctly human phenomenon—one that we had evolved to enhance our great intellect and unusually communicative nature. More recent research suggests that we are not as special as we thought. First of all, there is now debate about how clear-cut the asymmetry of many human brain functions really is—although language is still generally accepted to be very wonky. And the frontal lobes, so distinctively large in anxious, planning humans like us, show very few right-left differences at all. Second, asymmetry is now thought to be present in all mammals, and perhaps even all vertebrates. There have been detailed studies of changes in Yakovlevian torque during the evolution of primates—it can be calculated from the impressions left inside fossil skulls by the long-decomposed brain. But brain asymmetry takes many different forms, and it has been suggested that animals tend to concentrate on things that they are especially good at on one side—speaking in us, and singing in birds, perhaps.

But the final part of the asymmetry jigsaw is still missing—how does the brain get so lop-sided? This would be easy to explain if the left language regions grew as children learned to understand, then speak, then read, and then write, but they do not. Most of the asymmetry of the human brain is first established in the middle of fetal life, although it may be enhanced later. Brain asymmetry is built into us very early on—long before any of the lop-sided areas are used very much. In species that develop in eggs, such as birds, brain asymmetry has been claimed to result from the way that chicks lie twisted to one side, with one eye facing the shell and thus receiving more light. The idea that quantity of incoming information might skew the growth of the entire brain may seem fanciful, but exactly the same argument has also been put forward for humans, the majority of whom lie on one side of the maternal womb in late pregnancy. And do not forget that back in the mists of Chapter 13 we encountered an ancient and bizarre eye-like structure on the top of the interbrain that is genetically instructed to form on the left. It is worth repeating the speculation that this organ, the paraphysis or "parietal eye," might be the seed of greater asymmetries present in the modern human brain.

So we have now trod lightly through the fields of the mind, through the mysterious cortex about which we hear so much. Not such a mystery, really, just a recent mammalian invention: two convoluted sheets of nerve cells underlain by an interconnecting tangle of arcing fibers, dedicated to the strangely mismatched marriage of smell and our "higher functions." In a simplistic way, its back part can be seen as sensory (vision in the occipital lobe, hearing in the temporal, and touch in the parietal) and its front part as motor, but there is a great deal of spare space for all that tantalizing "association" to go on. And the whole cortex is lop-sided in many respects, but somehow the new corpus callosum helps to hold our two minds together. The cerebral cortex turned out to be not as much of a cerebral jungle as we feared. Instead it is a veritable springboard of ideas from which to launch ourselves toward comprehending intellect, understanding, and consciousness.

But we are not yet quite ready to make that leap. First, we must complete our tour with some deep, dark, dank recesses of the hemispheres. Let us take an emotional trip down memory lane.

18

THE SEAHORSE AND
THE ALMOND

Memory, Learning, and Fear

Imagine the scene. You are sitting on a park bench. The sun is shining and the dew-kissed grass is green around you. People are walking past and children are playing. All seems peaceful and as it should be—the world ticking past you.

Yet that world is badly wrong. A sudden fear grips you as you realize that something important has been pulled away from beneath you. Everyone else is happy to be getting on with their daily lives, but you are different from them. You remember nothing, absolutely nothing, about your life up to now. There is an infinite stretch of emptiness leading up to this moment—no past experience at all. Superficially you are able to function quite well—you can see and hear perfectly; you can move your limbs and face and you can rise and walk and run; you can understand what people are saying; you can talk and you can name all the things you see. But there is nothing of your past left in your head. You know that you are a person just like everyone else, but you have no idea who you are. You look down at your clothes for clues to your identity, but nothing brings the past flooding back. Nothing. With no memories your mind seems frighteningly rootless. You abruptly realize that you can remember things that have happened in the moments since you found yourself sitting on the bench, but that realization makes the loss of all that went before seem even more stark. Why did it go? Will it come back?

It is hard to imagine what this must be like. Sudden loss of all memories up to a single, watershed point in time is one of the rarest forms of amnesia, sometimes called the "fugue" state. The whole of life up to that point becomes a forgotten introduction to a disconnected flight into the future. Prelude and fugue.

Fugue amnesia is such a haunting idea that it often makes its way into the popular press. Someone simply appears somewhere in the world and no one, including the victim himself, knows who he is. Sometimes they have evidence of a head injury and sometimes they do not. They can move and speak normally, often retaining linguistic and musical skills from their past. They usually seem to be entirely conscious—and that is an important word—of their situation. Some are frustrated and frightened by their sudden severance from the past, but in others it drives an enthusiastic striving to appreciate the freshness of their "new" life.

Memory, fright, emotion, motivation. One can only guess at the flux that must be going on inside the brain of someone in fugue. Yet those four—memory, fright, emotion, motivation—are not now thought to be as esoteric as they might seem. As with perception, interpretation, and context before them, these four concepts have also been conquered by the cartographers of the brain. They now have a degree of definition and a sense of location that we could not have imagined a century ago. Are we edging closer to allocating every conceivable function of the mind to a specific patch of pale brown blancmange? And can that localization help us understand why some people have their emotional world or their past life taken away from them? A place for everything and everything in its place.

For many decades we have suspected that things like memory and emotion are located in a set of structures located deep inside the brain. Ever since anatomists first studied the cerebral hemispheres, they realized that there is something exceptional about their deepest parts. They seem tangled and distorted and altogether unlike the homogeneous convoluted sheet that coats the outer surface. Their unusual structure, often visible to the naked eye, meant that they were easy to find—standing out clearly against the uniform backdrop of the rest of the cerebrum. Because of this, these entities could be located easily not only in people, or mammals, but throughout the vertebrates. And when the microscope was discovered, they were found to differ fundamentally from the rest of the cortex, for example never possessing the usual six layers of neuron cells. They were a visible common theme running through all the different groups of back-boned animals, so when we started to understand the process of evolution, we decided that they must be very old. This collection of deep brain structures was given the name archicortex, or "old cortex": a name it is still often given today.

Although "archicortex" makes these deep structures sound like the elder statesmen of the brain, we now think this term is misleading. As you may remember, we have moved away from the idea that the parts of the forebrain that happen to be large in humans must therefore be newer and more advanced. Our evolutionary studies have shown that all vertebrates probably share the same basic arrangement of the forebrain, but that different regions are emphasized in different groups (see Figure 17.1). The deep structures we are discussing now correspond to the inner pallium—the regions on the inner surface of each forebrain bulge which face each other across the midline cleft that separates the two hemispheres. Thus, the archicortex is not unusually old—it is simply unusually easy to find in many different types of vertebrate.

So we need a different name for these bits and pieces involved in memory and emotion, but I must admit that we do not have a good one. Many medical neurologists and psychiatrists call this region the medial temporal lobe, which is at least a fairly accurate name. These structures are lodged on the inside surface of the large human temporal lobes. However, I feel that this name demotes them somewhat because they were present long before the cramped mammalian cortex had to fold around to form temporal lobes. The deep structures are a long, proud tradition in brain evolution whereas the temporal lobe is a Johnny-come-lately in comparison, so to name the former with reference to the latter seems disrespectful.

The other name often given to these structures is probably the one we must stick with, even though it is vague, misleading, and harks back to a time when we misinterpreted their role. I am sure that grouping them together as the limbic system confuses some into thinking that they have something to do with moving limbs. Yet here the word "limb" is used in its broadest sense—meaning "something at the edge." After all, your arms and legs are "at the edge" of your body in a rather unconvincing way. In a more convincing fashion, the deep memory and emotion regions of the hemispheres trace two bounding loops—one on the left and one on the right—around the periphery of the interbrain. And as with so many things in this book, the limbic system was given its cryptic name before we had much idea whence it came or what it does.

The distorted, looping pattern of the human limbic system is a result of its evolutionary history. Like the other parts of the forebrain, it has been with us a very long time, but it has been passed down to us in an unusually contorted form (Figure 18.1). In many vertebrates, the inner pallium probably acted as a simple linear conduit for smell information to pass from the two sides of the nose to the brain stem. These right and left pathways were important for allowing our forebears to sniff out things they liked and to avoid the scent of things they feared. At some point, they also developed the ability to remember where in the environment nice smells could usually be found. These paired inquisitive-smell pathways have remained with us to the present day, but their anatomical complexity has been forced upon them by a thoughtless neighbor.

As we saw recently, when mammals evolved their reorganized hemispheres, they also evolved a new link between those hemispheres—the corpus callosum, which we encountered in our brief look at the asymmetries of the cortex and the unusual world of split-brain patients. The callosum initially sprouted from the top of the interbrain, squeezing the right and left sides of the limbic system upward into an arch-shaped detour (see Figure 18.1b). As the callosum enlarged further, it spread backward, tugging the ever more circuitous limbic system with it, until those old inquisitive-smell pathways now have to effect a double arc to reach their eventual destination. First of all, a fold of cortex on the inner surface of each hemisphere called the cingulate gyrus—meaning "girdle"—arcs above the callosum, from the front of the brain to the back (Figure 18.1e). There it connects to a structure shaped like a curved sausage called the hippocampus. The hippocampus is a creature of mixed metaphors, however, and its front extremity is called its pes or "foot" because it has some deep furrows

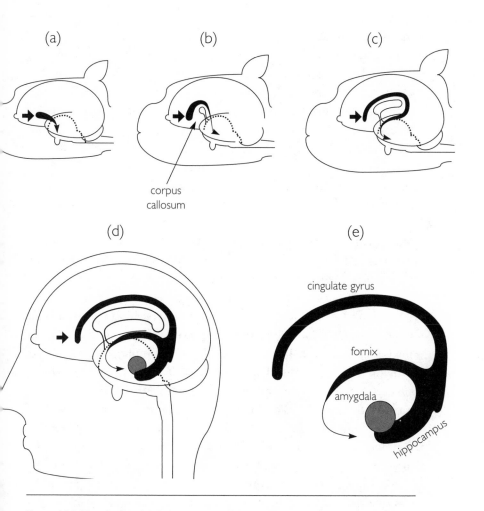

(a)   (b)   (c)

corpus
callosum

(d)   (e)

cingulate gyrus

fornix

amygdala

hippocampus

Figure 18.1. The distorted, looping pattern of the human limbic system is a result of its evolutionary history. (a) In many vertebrates, the inner pallium acted as a simple linear conduit for smell information to pass from the two sides of the nose to the brain stem. (b) As mammals evolved their cortex further, they also evolved a new link between those hemispheres, the corpus callosum. The callosum initially grew out of the top of the interbrain, squeezing the right and left sides of the limbic system upward into an arch-shaped detour. (c) As the callosum enlarged further, it spread backward, until the limbic-system pathways had to effect a double arc to reach their eventual destination. (d) The location of the limbic structures within the human brain. (e) The limbic system consists of paired left and right structures—the cingulate gyri arcing over the callosum, the hippocampi at the back, and the fornices arcing under the callosum. The amygdalae are perched on the "feet" of the hippocampi.

that almost appear to divide it into "toes." On this hippocampal foot sits the amygdala—an elongated blob, not unlike an American football perched on a kicker's toe (we will look at the origins of "hippocampus" and "amygdala" later).

From the hippocampus, a return pathway arcs underneath the callosum, returning to the front of the brain as the fornix. Finally the fornix loops backward on itself once more to at last reach the destination of the limbic system—the brain stem. I would love to be able to tell you that in this context "fornix" means "brothel," as most dictionaries will tell you, but I am disappointed to say that it merely means "arch." It is perhaps enlightening enough to know that the classical practice of ladies of the night soliciting for business beneath archways has meant that the outflow of the limbic system has common etymological origins with the word "fornication."

So the limbic system is an arching, recurved structure—a long detour around some of our newer evolutionary acquisitions. Like a lump of dough folded once, then twice, and then left on the slab, it is the limbic system that gives the inside of the brain the appearance of something kneaded. Its elegant elaboration has long made scientists suspect that it does something important. Eminent thinkers of the nineteenth century, including the influential paleontologist Richard Owen, thought it was central to consciousness, self, the soul, and humanness. It was even claimed that it had a subsection unique to humans, the hippocampus minor—the physical cause and embodiment of our superiority and stewardship over the beasts. And when biologists started to study the effects of damage to the limbic system, they were tantalized by the profound and interesting effects they observed. Sometimes damage caused an apparent flattening of emotional responses, even signs of passivity or apathy, although alternatively it could lead to apparently random behavior in which tasks were carried out in an entirely unpredictable order. By the start of the twentieth century, these effects had led to the grand idea that the limbic system was involved in emotion and motivation, and thus might be the key to understanding our selves.

However, both emotion and motivation were poorly understood phenomena at that time, and the same remains true today. Biologists brought up in the shadow of Darwin try to relate the things they study to the ways in which they might give animals an advantage.

Take emotion for example—why do people and animals do it? In fact, there are several theories that attempt to explain why animals find emotions useful, and many of them are based on the idea that we need emotions to drive our actions. For example, it was long thought that once the conscious part of our brain has analyzed a situation, then that analyzed information is fed to a deeper, emotional part of the brain. This emotional brain then drives our responses to the situation—emotion as a spur to motivation. A contrary theory was that we make instinctive responses to certain situations, and that it is our own responses that cause our emotional reaction, rather than the situation itself. So maybe little children "make themselves cry" to induce the emotional state of "sadness." A third, and very different theory, is equally good at explaining the little child's actions. Perhaps we have emotions so we can communicate our state of mind to other people. After all, we usually speak of "showing" emotions, rather than simply "experiencing" them. Maybe emotions are a communicative tool—do we show fear to elicit protection from others, happiness to promote integration into social groups, and anger to emphasize our feelings?

All these conflicting explanations for emotion—driving our actions, being driven by our actions, and as a form of communication—have probably confused our search for the location of emotion. By the 1930s, it was thought that events in the outside world could affect two areas of the brain—one, probably the hypothalamus, involved in driving responses to those events and another, perhaps the hippocampus, forming the emotions themselves. And in 1937 a neuroscientist named James Papez proposed the existence of an emotional "circuit" inside the brain around which these responses could circulate, occasionally throwing off impulses to drive bodily responses as well as conscious emotions. The Papez circuit looped and somersaulted through the cingulate gyri, the hippocampi, the fornices, the "breast-like" mammillary bodies, those verbose mammilothalamic tracts of Félix Vicq d'Azyr, the thalami, and then back to the cingulate gyri. The circuit was thought to circulate and recirculate impulses, driving feelings and responses until the impulses were spent. Like some emotional washing machine, the circuit tumbled and mixed our feelings until they were ready.

The Papez circuit was pleasing for all sorts of reasons. First of all, although circuitous, it was clearly defined and it was easy to see where sensory information could feed in and emotional responses could be output. Second, it unified the otherwise rag-bag assortment of things in the limbic system into a functional unit. And third, it chimed nicely with the idea that it had evolved from an older, simpler system used by our mud-grubbing aquatic ancestors to sniff out the things they needed and things they feared. But times have changed. It is hard to tell whether the Papez circuit formed the basis for our modern understanding of emotion, or whether its progressive adaptation has left the original theory in tatters. It probably depends on how you think emotion, memory, and motivation relate to each other. Some parts of the circuit now seem to have functions unrelated to emotion, while emotional roles have been ascribed to structures not included in the original circuit. The foremost of these are the amygdalae.

You have two amygdalae in your head—one on the left and one on the right—and as we have seen they are perched on the front tips of your hippocampal "feet," a few inches in from your ear (see Figure 18.1d,e). They are elongated blobs of grey matter similar in size and shape to almonds, which is what "amygdala" means. However, their internal structure is complex, and continuing anatomical study has led to ever more cumbersome names being applied to their subdivisions—including, for example, the snappy "nucleus amygdalae basalis pars lateralis." They are present in all vertebrates, but their origins are unclear. In fact, they may be composite structures—a mixture of a lump of grey that always lay deep inside the hemispheres and some fragments of cortex that became internalized at a later date. Confusingly, some of the bits that are noncortical in origin can function in a way similar to the cortex, and are sometimes called "vicarious cortex." So the amygdalae are truly strange and, as we will see, frightening places.

The amygdalae are difficult to study, which probably explains why we did not discover their functions sooner. They are buried deep beneath other fragile, important structures, and so it is almost impossible for people to suffer an injury that affects both their amygdalae and nothing else. Because of this, much of what we know about the amygdalae comes from experiments on animals. When the amygdalae are destroyed, adult animals become extremely tame, and they also interact with their fellows a great deal more—it is as if their fear and inhibitions have been removed. Similar changes occur in the rare condition Urbach-Wiethe disease, in which the human amygdalae progressively degenerate. These patients lose the ability to experience anger and fear, and even the ability to recognize the emotional importance of facial expressions in others—remember the special eye-hillock-almond pathway for fear recognition I mentioned in Chapter 11?

But the amygdalae are more complex than simply being the place where fear is recognized. They receive many inputs, the largest of which are, perhaps unsurprisingly, from the association cortex—in fear, context is everything. They also gather smell information from the nose and the vomeronasal organ, maybe as a relic of an original mud-sniffing function. Their main added layer of complexity is that they do not simply respond to frightening stimuli, but they are also the place where fear is learned. Fear seems to have its own little place where it is committed to memory, and this exclusivity may explain why memory of fear can seem so obscure. Stress, anxiety, phobias, and post-traumatic stress disorder are all examples of conditions in which sufferers can logically see that their fears are unhelpful, but they cannot seem to use their conscious mind to defeat those fears. Thus, the inaccessibility of the amygdalae to conscious perusal may explain many of the ills of modern life.

The hidden tangle of the amygdalae sends even more tendrils out into the world of our suffering. There are differences in the gross structure and size of the amygdalae in people with depression, people with a predisposition to alcoholism, as well as differences between the ways that men's and women's amygdalae react to frightening things. Little flurries of electrical activity in the amygdalae, akin to tiny bursts of epilepsy, have been implicated in unpredictable outbursts of rage and violence. Damage to the amygdalae of young primates induces a state that superficially resembles autism, in which the infants do not play with their peers but instead carry out repetitive, ritualistic behaviors. This is not true autism, however, as these individuals are very alert to what is going on around them. Thus, the role of the amygdalae in our normal emotional development is a complex one, and one that also involves our appreciation of pleasant as well as unpleasant stimuli in the world around us. Most worryingly of all, development of the fetal amygdalae may be influenced by maternal hormones, which has led some researchers to suspect that maternal stress during pregnancy may predispose children to attention deficit disorders and depression later in life.

The profound and varied roles of the amygdalae have explained a great deal, but the almonds have in some ways rather hijacked our investigation of emotion. Everything has become very negative—the adaptation of a system that detected unpleasant smells to new roles in fright, anxiety, and violence. Where is the positive side to all this? Which parts of the brain counter the amygdalae? After all, many of the decisions we make are positive. Motivation is often a positive thing—we do things because we want something; we do things to achieve the warm glow of success for its own sake; we do things because it feels good to emulate our parents, mentors, or heroes. We climb mountains because they are there. The fearful amygdalae have snatched away much of what we thought the rest of the limbic system was doing. What, for example, have the hippocampi been left to do? As we will now see, they have taken up a central role in memory, and hopefully I can convince you that it is memory that can, perhaps unexpectedly, be seen as the positive side of all that almond-scented negativity.

The hippocampus is bigger than the amygdala, and it twists around the depths of the back of the brain. You may remember that I described it as being shaped like a curved sausage, but fortunately it is instead named after its appearance when it is sliced through. The hippocampi form as infoldings of the inner sides of the cerebral hemispheres—they are really just a special type of sulcus (Figure 17.3). They fold in a complex and sinuous way, however, and in cross section the layers of cells have an appearance not unlike the graceful profile of a seahorse, and this is what "hippocampus" means.

Much of the hippocampi comes to lie below the corpus callosum as it forces its way backward, although a small strand of hippocampal tissue, the induseum griseum ("grey undergarments") of Lancisi is stranded above the callosum. The hippocampi each connect at the front to the fornices, which communicate left to right via a band of fibers called the commissure of the fornix, also more ecclesiastically known as the psalterium. The internal folded structure of the hippocampi is even more complex and consists of several portions—the dentate ("tooth-like") gyrus, the alveus (or "river channel"), the subiculum ("support"), and the horn of Ammon (later depictions of the ancient Egyptian god of the air, Amun, represented him as a curly-horned ram).

This anatomical mixture of mythology, church, and underpants is all very well, but are two curved sausages really sufficient to store all our memories? There are certainly a few problems with that suggestion. First of all, we have already seen that fear is remembered somewhere else, setting the precedent that different types of memory might be located in different places. Also, memory is a notoriously capricious thing—we use it all the time, but if often fails us for no good reason, and we are never entirely sure how reliable our memories are. Finally, although Penfield could induce memories by stimulating certain cortical regions, the destruction of no single patch of brain wipes out all our memories. So perhaps if we want to find out where our memories are stored and how the hippocampus helps us with that storage, we should think first about what sorts of memories we have.

First of all, we can forget about fear, as it has its own storage site. However, there are some other simple types of memory and learning that may have their own special locations. Motor skills like walking or playing a musical instrument are probably filed away in the basal nuclei and cerebellum. Also, we have seen that language skills are stored in special areas of the (usually left) cortex. These different types of memory probably explain why fugue amnesiacs can walk, talk, fear, and play musical instruments, even though they cannot remember their own name.

Another low-level form of memory and learning will forever be associated with the Russian experimental biologist Ivan Petrovich Pavlov. Pavlov was actually interested in the mechanisms of digestion, but during his studies he fortuitously discovered a mechanism by which animals learn. When an animal sees an item of food, it salivates in anticipation of eating. Of course, this is entirely sensible—it is good to get the juices flowing before a meal. If the appearance of that food is preceded by some other stimulus, for example, ringing a bell, the dog will still salivate. This may not surprise you, but what Pavlov

discovered next was fascinating. After a few bell and food presenta-
tions, if the bell is presented alone, the dog will still salivate, even
though there is no food offered. So Pavlovian conditioning consists of
a sensible response transferred to an arbitrary stimulus by a process
of training. This is actually far more important than it sounds, because the responses are not restricted to simple things like salivation,
and people can be conditioned as easily as animals. The rules are very
strict, however—during training, the arbitrary stimulus must precede
the sensible one, and the time gap between them must be short. This
form of learning may seem alarmingly crude, but it is extremely powerful. You may be able to think of some examples in which an abstract
stimulus induces an entirely illogical response in you. For example,
the linking of responses to inappropriate stimuli has been suggested
to underlie phenomena such as phobia, fetishes, and some other be-
havioral idiosyncrasies.

But when most people think of memory, they do not think of language, walking, or fear, or the way their heart races when the telephone rings late at night. Instead, they think of the everyday things
they experience and later consciously recall. Information probably
flutters though various different states as it makes its way from our
sensory perceptions to our long-term memory banks, and then back
to be remembered, but there seem to be two very important stages
through which it must pass. The first is often called short-term memory, where small, recent snippets of experience are held and then
used, stored, or discarded. Your short-term memory is probably very
low in capacity, and it is really quite easy to devise tests to measure this capacity—if you experiment, for example, you will find
that you can reliably remember six-digit numbers for short periods,
but nine-digit numbers overload your short-term memory. The other
important thing about short-term memory is that it might be multi-
purpose—it may also be used as a temporary workspace to handle information retrieved from your other memory banks.

Of course, long-term memory is the persistent, high-capacity data
storage facility in your head. It is much harder to assess the size of
your long-term memory as it contains some information that you can
recall easily and other information that you can only remember after
being reminded. Also, long-term memory is probably more textured

than the other types of memory. As information is committed to long-term memory, the actual perceptions begin to fade and are replaced by interpretations of those perceptions. Meaning starts to dominate sensation, and when those memories are recalled it is likely that we must reconstruct the sensations from what we remember of their significance. In our mind's eye, we idealize our lovers and demonize our enemies—or is it the other way round? Not only that, but every memory is picked apart into its constituent components, and each is stored in a different place in the cortex. Names, colors, textures, movement, location, time, and context are scattered among the convolutions. This scattered nature of long-term memory is why no single injury can selectively obliterate just one memory. Instead, head injuries can expunge the same aspect of many memories—for example destroying the ability to draw remembered things.

Startlingly, these concepts of short-term and long-term memory are reflected in the different types of memory loss that people suffer. For example, fugue amnesiacs lose access to their previous long-term memory, but they can form short-term memories and transfer them to make new long-term memories. More commonly, people lose the short-term memories they acquired immediately before an accident or frightening episode, but their memory often functions normally after the accident and they also retain their long-term memories from before it. Long-term alcohol abuse can damage the process by which short-term memories are committed to long-term memory, and these patients drift through an unnerving short-term world, even though they remember everything from before the damage was done. And the scattered nature of long-term memory is dramatically supported by the cases of amnesiacs who can identify iconic figures such as Winston Churchill and Che Guevara but do not recognize members of their own family.

An especially important insight into mechanisms of memory came when a human patient, H.M., had both of his hippocampi deliberately removed in an operation. This man has become something of a neurological celebrity and access to him is closely guarded by a coterie of protective neurosurgeons. What is remarkable about this individual is how he retains his long-term memories from before the surgery, can form short-term memories, but is no longer able to transfer them to his long-term memory. He travels though life in a moment-to-moment fashion, still believing that he is living at the moment in history when he entered the operating theatre—now some decades in the past. What this shows is that the hippocampus is not the site of actual memory storage, but is instead the structure that transfers memories from the transient short-term workspace to the long-term memory filing cabinet.

Now that we know where to look, the enigma of memory is slowly unfolding. Millions of neural pathways are known to snake their way through the folds and arcs of the hippocampus in a highly ordered manner. Information flows into the hippocampus from the surrounding areas of cortex and is either marked for safekeeping in other regions of cortex or consigned to the memory dustbin. The hippocampus is the master controller of memory. Of course, this disposal is extremely important because your memory has to be highly selective—how much of what happens during a day do you actually need to remember? It has even proved possible to identify some of the microscopic and molecular changes that occur when memories are formed. One of these changes, called long-term potentiation, occurs when a neuron can be rapidly tickled into weeks (at least) of excitability by an apparently innocuous input. It is, in effect, remembering that input. And our new toolkit of molecular biology is allowing us to identify the molecules in nerve cells that actually effect this memory. We can even remove slices of living hippocampi and keep them alive in a dish, and under these alien conditions their cells can still industriously undergo long-term potentiation. A slightly less unsettling line of research has shown that London taxi drivers grow their hippocampi as they learn the road layout of the city.

So the seahorses and their arcing connections are a central switchboard for memory. But where does that leave the old idea that the limbic system is a coherent unit within the larger scheme of the brain? If the hippocampi process memory and the amygdalae process fear, how can these two be reconciled into a unified whole? The whole idea of the limbic system is often seen as rather shaky these days, with a seemingly random assortment of cerebral bits and pieces included within it or excluded from it, depending on which neuroscientist you ask. However, I would argue that if we look back at the evolution of these tangled structures at the center of our heads, there is a story there to be told. If we return to our aquatic ancestors swimming about in the primordial gloom, we can surmise that they needed to respond to two things.

First of all, dangerous things were inherently unpredictable and were often detected by their smell—so the amygdalae developed to remember what frightening things smell like and to react to them with a protective response called "fear." Obviously other senses could feed into the amygdalae as well, and these senses have come to predominate in humans, but the principle remains the same. The second thing that our ancestors had to detect was attractive food items, and presumably they once again did this by smell. However, in many cases these food items could often reliably be found at certain locations within the animals' foraging range. Thus the hippocampus developed an ability to remember not only what food items smell like, but also a spatial map of where that food may be found. Many vertebrates retain this system today—fish, birds, and mammals, at least—using their hippocampi to develop an accurate spatial map of their surroundings to help them forage and migrate. The hippocampus is, for example, charmingly large in many species that store food in scattered hiding places.

So the hippocampus is there to remember things we like and the amygdala is there for things we do not. Overly simple, perhaps, but at least it is a theory that tries to cut through the confusion. Many neuroscientists do now believe that memory and emotion, while seeming subjectively quite different, may have a great deal in common. Certainly, detailed studies of the molecular mechanisms of the two are revealing unexpected similarities. But all this talk of memory and emotion leads us on to yet more questions. If we can localize memory and emotion, then are we not nearing the end of the list of things we had to find in the brain? And if we can understand memory and emotion, is that not a great step to understanding consciousness itself? When I first planned this book, I eschewed the idea of writing yet another book about how the brain works, but I have finally come to the stage where I can no longer avoid the difficult issue of consciousness.

After all, it would be churlish not to have a go at it.

19

THE HARD QUESTION

Brain Size and Consciousness

Consciousness is not explained to my comprehension by all the nerve-paths and neurones of the physiologist; nor do I ask physics how goodness shines in one man's face, and evil betrays itself in another.

—D'Arcy Thompson, *On Growth and Form*

I remember Sunday the tenth of August 2003 very well. It was the hottest day ever recorded in Britain, with the mercury edging over 100 degrees Fahrenheit for the first time.

Of course, on this furnace of a day it fell to me to be on emergency call for the local veterinary practice. The previous week had seen the temperature gradually rise so that animals were succumbing to heat stroke in their usual preordained order—the feeble dogs and rabbits were the first to expire, and by the weekend, even those with a stronger constitution, the guinea pigs and the rats, were starting to drop. In fact, by the Sunday itself things had become quieter, presumably because most potential heatstroke victims were already dead. Then, at three o'clock in the afternoon, I received a call about a cat that had been accidentally shut in a car since eight o'clock that morning. I mentally groaned as I envisaged having to end the suffering of some poor creature whose internal organs had disintegrated in the heat. However, when I asked the owner about the state of said feline, they replied, "Well, he was alright, but we tried to dunk him in cold water and that made him *extremely* cross."

That, in a nutshell, is cats' attitude to extremely high temperatures. They are designed for them, and often even seek them out. Probably descended from the African desert cat, they know to lie low in the shade until the heat is off. They do not pant, or gasp, or panic: they just look a bit angry. Like many desert species, cats have a meshwork of tiny arteries underneath their eyeballs that supply blood to the brain, possibly cooling it by routing it alongside the chilled blood draining back from the nose. This arterial mesh is the *rete mirabile* or "wonderful net" that Galen discovered in so many of his specimens— many domestic species have retia under their brains or eyes. Galen was fascinated by the retia, but it was probably some of his more overzealous disciples who claimed that they were the location of the soul. As we saw in Chapter 2, this was rather ironic, as the retia are one of the very few brain structures that we do not, in fact, share with the beasts. Thus this theory would consign us to a soulless place in the order of things.

What this does show, of course, is that classical antiquity was a time when people were happy to accept that animals had a "soul" just as we do (there was no separate word for "consciousness" in the Western tradition until the eighteenth century) and that it was widely believed that there was a specific part of the brain where that soul might be found. Truly a golden age of enlightenment.

But things seem more tangled now. The more we thought about consciousness, the less it seemed compatible with the messy world of the flesh. In the minds of later philosophers, the soul and the body were gradually teased apart until Descartes conceived of the soul guiding the body by tenuous reins—controlling muscles and sinews via the tiny pineal orchestrator atop the brain. And as our understanding of the brain improved into the twentieth century, and we were able to discern and localize ever more complex functions to various parts of the brain, things became, if anything, yet more obscure. Defining brain regions that see, hear, interpret, contextualize, plan, and move us served only to make the "hard question" of consciousness seem harder still. All those higher-level functions of the brain were becoming definable, tangible, measurable, and objective, but consciousness was none of those things.

We all have a strong subjective feel for what our own consciousness is, but it is just that: subjective. We cannot measure it in any way, nor define it with respect to anything else. And despite its overarching importance to each of us, it is unique in that it is absolutely personal. We all have complete access to our own consciousness, and indeed our own consciousness may in some way "be" us, but we have absolutely no access to the consciousness of anyone else. We have already confronted the irony of how our most compulsively inquisitive and social organ is also our most physically imprisoned organ. Now we must come to terms with the fact that the function of our brain that seems to rise above the mêlée and lets us actually know ourselves is entirely cut off from other individuals. Every one of us is an island, after all?

Yet all this bemoaning our eternal isolation from each other will not help. What we need to understand is whether or not there is something fundamentally unknowable about consciousness. Should we set it above all the other phenomena in the universe and accept that it is something to which science has no access? You may, of course, already have your own opinions on this matter, and certainly there is a whole spectrum of possibilities. At one end of the spectrum there is the possibility that consciousness is so inherently personal that it can only be studied by subjective self-analysis. At the other end is the idea, sometimes called the radical neuron doctrine, that everything in the mind can be explained by studying the interactions of nerve cells and that these studies will eventually supersede all other theories about the brain. I do not want to prejudge this issue, largely because that would mislead you into thinking that I am going to give you a simple answer at the end of this book. Instead I would like to start out somewhere you might not expect. I would like to start our final dance through the fields of the mind by thinking about other animals.

Much of what we know about consciousness and the brain comes from animals, which raises a big question—are animals conscious in the same way as us? Obviously, we cannot answer this when as yet we have no definition of consciousness and certainly no objective way to measure it in others. But I think it is good to start with a question that places people in context. For 3 billion years the world was covered in life, but there were no people. And then for the next short few million years we humans have been here. Was there really no consciousness until this last flickering instant of cosmic history? No animal able to make even the most simple assessment of its own existence? I personally believe that my cat is as conscious as I am, but do I really have any justification for that belief?

I hope that one thing of which this book has convinced you is that there is nothing qualitatively unique about the structure of the human brain. Yes, it is big and some of its parts are disproportionately so, but there is no one novel chunk of it that stands out as definitively human. The only regions that one could realistically claim are unique are the language centers of the left hemisphere, but we will return to those later. That possibility aside, there is no human-defining brain region, no "nucleus humanus," no analogue of Owen's hippocampus minor. In short, there is nothing in the blueprints that tells us we are special.

And yet there is one feature of the brain that human supremacists consistently champion—its size. The story of the recent evolution of the human brain is a story of a spectacular increase in brain size, perhaps unprecedented in all the annals of biology. Five million years ago, our ancestors' brains were small—maybe a third the size of ours—and 5 million years is a very short time in evolution. You often hear the statistic that our closest cousins the chimpanzees share 99 percent of our genes. Well, one percent is still a lot of genes, and certainly more than enough to make some serious differences to the brain. Clearly, there is something inherently important about human brain size, but is it the defining feature that places our species above all others? In short, have we and we alone exceeded some critical mass for the brain, some threshold size above which we are able to do fundamentally new and unique things? Does a brain suddenly flicker into consciousness when it swells past a certain level? It would certainly be intriguing to link that most highfalutin' concept of mind—consciousness—to its most base measurement, its poundage.

But there is more than one way to look at the weight of brains. The main reason for this is that as animals get bigger, they simply seem to need bigger brains. A whale's tail has maybe a million times more muscle cells to move it than a mouse's tail, and that discrepancy requires that there are more nerve cells required to control those muscle cells. Maybe not a million times more, but certainly more. Whales could have countered this problem by having fewer but bigger muscle cells, but there are mechanical and electrical problems with doing that, so they still need many more nerve cells to control their bodies. Another thing that tends to make large animals' brains bigger is that they have bigger neurons. Not vastly bigger as, again, this might compromise their function, but they do seem to have swollen somewhat in animals where space is not at such a premium. You might argue that, unlike the parts of the brain that control muscles, the regions that interpret sensory information would not have to be larger in big animals—after all, a noise or a smell is a noise or a smell whatever size you are. However, for some reason even the sensory regions are larger in big animals.

So if sheer size is our sole criterion, then the cleverest and presumably most conscious animal is the sperm whale with a brain of 8,000 grams, compared to bottlenose dolphins at 1,600, humans at 1,300, and chimps at 500. With all that cogitation and self-analysis going on, the oceans must be like some Parisian Rive Gauche café of the 'forties or 'fifties. However, if bigger animals simply get a bigger brain because they are bigger, then perhaps it is not justified to compare cetaceans to Camus. Instead, a more sensible thing might be to calculate the fraction of body weight that is made up of brain. If this is done, then things look a little more anthropocentrically pleasing. Cows' brains are one-thousandth of their weight, dolphins' are one-hundredth, and humans' are one-fiftieth. This may seem just about right to you, all things considered, but I have some bad news. Mouse brains are one-thirtieth of their body weight and small birds can be as much as one-twelfth—relatively far larger than ours. And as you might imagine, the sperm whale does not fare very well at all.

The reason for these apparent inconsistencies is that, although as animals get larger their brains also get larger, their brains do not enlarge as rapidly as their bodies do. A whale may be a million times larger than a mouse, but its brain is nothing like a million times larger than a mouse's. Brains do not have to keep pace with bodies, and this means that larger animals' brains are actually relatively smaller. In fact, scientists have now discovered something startling—that you can use some simple arithmetic to link brain and body size. Why the natural world should obey the laws of arithmetic, you may wish to consider, but it does. For any group of animal species, we can predict for each species what size its brain should be simply by placing the animal on a weighing scale. And for each type of vertebrate, this system works remarkably well. We have a formula for birds, for example, that works for hummingbirds, seagulls, and ostriches. For fish we have a formula that works for guppies and tuna. Yet there are noticeable trends among the major groups. Mammals have slightly larger brains, on average, than similarly sized birds, and they in turn exceed those of equivalent reptiles. There is considerable overlap, but the trends are clear.

There are some equally intriguing trends within the groups as well—some animals seem to wander from their predicted brain size in a noticeable way. Active, predatory fish have larger brains than expected. Fish with one unusually well-developed sense, such as electrosensitive mormyrid "elephant" fish, have bigger brains. Birds who use tools, such as crows, and cetaceans with complex communication have unexpectedly larger brains. Domesticated animals have smaller brains than their wild relatives. And one of the most dramatic deviants is the human race—our brain exceeds the formula's prediction by an unusually wide margin. If we assume that a chimp's brain is exactly the size we would expect, then a dolphin's brain is twice what we would expect and a human's is three times. Considering how well the arithmetic works across such a large range of species, the unexpectedly large human brain really does stand out as unusual. Perhaps size is important.

Very recently, scientists have claimed to have discovered the exact genes that are responsible for making our brains so large. Some unfortunate children are born with a condition called microcephaly, in which their brain is unusually small—perhaps only a third of its normal size. Microcephaly may have many causes, but some cases have been linked to defects in just two or three different genes. The finding that genetic damage to these genes can cause human brains to revert back to a size similar to that of a chimpanzee has made us wonder whether it was modifications of these few genes that drove our acquisition of large brains over the last few million years. That a few genes could control such spectacular changes may seem strange but it is not unprecedented—there are other organs that have a single gene driving their entire formation. If this theory is true, microcephaly, although a tragedy, can be seen as an intriguing reversion, an atavism to our ancestral state.

The nineteenth-century German zoologist Ernst Haeckel suggested that the convolutions of the cerebral cortex could also be used as a measurement of intelligence: "In all human individuals distinguished by peculiar ability and great intellect, these swellings and furrows on the surface of the great hemispheres exhibit a much greater development than in common average men; while in the latter, again, they are more developed than in Cretins and others of unusually feeble intellect." This less than politically correct statement was in fact first made by the ancient Greeks. Humans have very convoluted cortices, whereas most mammals have more smooth, or lissencephalic, hemispheres. Like us, dolphins and whales also have extremely crinkly brains. Yet there is an over-simplification in Haeckel's statement. As we have seen, the cortex is essentially a sheet of grey matter coating the outside surface of the mass of the brain. But over two thousand years ago, Euclid showed us that as an object gets bigger, its surface area does not keep pace with its volume. Thus, if a brain becomes eight times heavier over the course of evolution, its surface area becomes only four times greater. So large brains like ours and cetaceans' are faced with the problem that they do not have enough surface area onto which they can fit all their cortex. To solve this problem they increase the area of the cortex by corrugating it—we saw how two-thirds of the human cortex is buried in the grooves and only one-third is on the surface. So corrugation is an almost inescapable result of brain enlargement, but not in itself a sign of intelligence. Of course Galen suspected this all those centuries ago when he noticed that donkeys have convoluted brains.

There are still some interesting deviations from this rule, however. First of all, social species often have more convoluted hemispheres than one might expect, suggesting that social interactions require a relatively extensive cortex. But my favorite exceptions to the rule are the spiny anteaters *Tachyglossus* and *Zaglossus*—a few species of endearing insectivorous, egg-laying mammal from Australia and New Guinea (a new species was discovered during the writing of this book). Spiny anteaters have surprisingly large and convoluted brains when one considers their simple earth-shoveling, termite-grubbing lifestyle. So why do they need all that formidable cerebral capacity? One charming theory suggests that spiny anteaters cannot dream, and so unlike other mammals they do not undergo an ordered process of discarding and filing their memories. Because of this, years of useless termite-related memories accumulate and fill their nonselective minds. A spiny anteater's sense of nostalgia must be overpowering.

So we can now see that even if there are no distinctively human regions in our brains, we are at least quantitatively different from other animals. By most sensible systems of measurement, we have very big brains, and we like to think that this makes us relatively clever. Whether or not it also makes us unique in experiencing consciousness is another matter, however. The human brain is nearly three pounds in weight, seven tenths of which is the hemispheres. It may contain 100 billion nerve cells with perhaps 100 trillion connections between them, but these are only estimates. The brain gets smaller in old age and following prolonged alcohol abuse. The brain is also horrendously demanding—although it is only one-thirtieth of your body weight, it probably uses one-fifth or one-quarter of your energy (some fish are even more brain-oriented than us, burning two-thirds of their energy in their brains). And strikingly, molecular activity in the human brain may exceed that of chimp brains by a factor of five.

And, as men love to remind women, their brains are larger: averaging 1,305 and 1,220 grams in the two sexes. This is a significant difference, but it is less than the difference in the average weight of the whole body, so women's brains are relatively bigger. There are also regions of cortex where women seem to cram in more neurons. People have tried to draw conclusions from these differences, but there is little consensus. Some studies show that the sexes differ in their performance and certain types of test—women having more verbal skills and men more visuo-spatial skills, for example—but even these widely reported differences are debated. Men often seem to be more variable in their abilities, with males disproportionately populating the "underachiever" and "overachiever" categories, but how this could relate to crude brain size is unclear. Controversial attempts to make similar generalizations about human races have foundered upon the interpretation of the results. After all, there are many subtle ways in which the races differ physically, so it is hardly surprising that one can design an intelligence test that favors one over another.

But before we leave the question of brain size, there is one more aspect which sets humans apart from almost every other animal, and that is the way the human brain develops. And much as I hate to admit it, there are some features of human brain development that are unique. At birth the human brain is roughly a third of its adult weight, 380 grams in girls, 400 in boys. This is similar to the brain of an adult chimpanzee. After birth, a baby's brain doubles in size in a year and has tripled by the age of six. A human baby probably forms a million new neurons every twenty seconds. This phenomenal growth is extremely unusual because in most mammals, brain growth slows after birth. In contrast, when plotted on a graph, a baby's brain carries on growing as if it were still a fetus, leading to the remarkable hypothesis that humans actually give birth to offspring that are only halfway through their fetal development.

This may all seem very spectacular, but I would urge a note of caution before we get carried away with the ferociously cerebral human child. Their unusual growth pattern can be explained largely by the simple fact that our brain must grow to be unusually large. And babies could probably be born with larger brains were it not for the simple fact that they do indeed need to be born. The newborn's head cannot be any bigger because it has to pass through its mother's pelvis—and women already have a uniquely inefficient, wiggling gait because of the recent re-engineering of their pelvis to allow their brainy offspring to escape. Of course our distant, communicative cousins, the whales and dolphins, do not face this problem because they do not really have much of a pelvis left. So our brain grows quickly after birth simply because it has to squeeze through a narrow tube and then must end up big.

The pattern of human brain development is more atypical than that, however. New evidence from modern imaging techniques has shown that some major elements of human brain development are not complete until the end of puberty, or possibly until the age of twenty-one. All the main pathways in the brain are established long before puberty, but much of their electrical insulation is not yet in place so they probably do not function very well. Some pathways and centers mature before others do, and this may explain why teenagers behave differently from adults. To oversimplify somewhat, during the second decade of life there is a gradual transfer of function from the emotional amygdala to the analytical prefrontal cortex. Is this why teenagers seem so overemotional and thoughtless to their bewildered parents?

So what have we learned so far? We have seen that human brains are quantitatively but not qualitatively different in structure and that this is largely the result of their unusual but not inexplicable patterns of development. We wonder if there is a threshold of size, or maybe relative size, above which brains become conscious. But as we cannot detect consciousness in another person, let alone an animal, we suspect that this may be idle speculation. And recent studies of children and adolescents have suggested that the nature of our consciousness may change considerably in our first twenty years of life. However, we have traveled through this book in the spirit of geographical explorers, and as yet we have discovered nothing about the location of consciousness. This is all proving very difficult.

How can we get some sort of grip on this evasive thing called consciousness? We all think we know what it is, and yet we can neither define it nor share it. We do not even know if we all experience it in the same way. Perhaps we should approach this strange phantom that flutters over our experiences, memories, and actions in as pragmatic a way as possible.

We need to address the assumption that consciousness is one of the things that our brain does. Not everyone believes that it can be explained in terms of brain activity alone, but I do, and so do most neuroscientists. If you believe that consciousness has part of its basis outside the brain, then that is mysticism and you will have to accept that science cannot explain it. Anyway, by accepting that consciousness is just one of the many things that the brain does, we have de-

moted it to a less esoteric level. For example, it must then obey the
same rules that other brain functions obey. If consciousness is part
of the brain, then we can answer the question "What are brains for?"
knowing that our answer will encompass consciousness. And certainly this question seems simpler than "What and where is consciousness?" Most scientists are Darwinians nowadays, so our question is rephrased as, "Why do brains help animals survive and
reproduce?" As we have seen, brains are unusually costly things, so
they must be doing something very useful to earn their keep. Evolutionary theories about brains vary, but they all have a great deal in
common. I would suggest that brains benefit animals by allowing
them to make appropriate responses to the environment. Others have
suggested that brains evolved to allow animals to cope with unpredictable variations in the environment. These two suggestions are not
identical, but they sound pretty similar—the brain as an internal re-
flector of, and responder to, the world.

The idea that our wonderful brain is simply there to respond to
the outside world may seem simplistic, but there is a lot to be said
for it. For wild animals, the world is a cutthroat place and the ability
to respond to change can easily make the difference between life
and death. Often what is needed are simple, quick, unthinking responses, and so most of what we do is unconscious. Even supposedly
highly evolved animals like dogs and people can undergo Pavlovian
conditioning so that they reliably and unconsciously make trained responses. Abnormalities in this process may cause counterproductive
and self-defeating behavior in confused modern humans, but these
simpler mechanisms of our brain usually work remarkably well and
save us a great deal of trouble. And there is no reason to think that
consciousness should be any different. We must assume that we have
it because it is useful for something—that it allows us to do something better than if we were entirely unconscious automatons. As we
stumble our way through the world of consciousness, bear in mind
this belief that consciousness must be of some use.

There are, of course, less utilitarian ways of thinking about consciousness. Philosophers love to play with the almost solipsistic idea
that were it not for conscious beings, the universe would have no observers and thus might not truly exist. That is all very well, but not

exactly relevant to this book. Scientists, on the other hand, have the more practical worry of what is the best way to investigate this thing called consciousness. As an anatomist, I would love to say that it can be elucidated by detailed study of all the interconnections of the cells in the brain, but I fear that this is not the best way to proceed. Not only would that involve manipulating a formidably unwieldy mass of information, but also it must be remembered that the brain was not especially designed so that its wiring would be easy for us to understand. Alternatively, many thinkers have tried to model brain processes using statistics, and this has yielded some general results, but there is a constant worry that the brain acts to emphasize interesting inconsistencies and informational "noise," rather than to smooth over them as statistical analysis does. In addition, over the last few decades the ever-increasing complexity of computers has made it a very real possibility that our best way to access consciousness is by empirically trying to create it inside a computer. Yet once again, there may be problems—for example, a designed mind may show fundamental differences from one that has evolved by unguided evolution—but a conscious machine would still represent a huge leap forward in our understanding of our own consciousness.

There are many different proposed definitions of consciousness. Although none of them is perfect, and some of them are very obviously imperfect, each of them raises different possibilities about how we think and how humans relate to the rest of the universe. Rather than being satisfyingly general and all-encompassing, each focuses on a different aspect of the brain's work, and perhaps that is their common failing. Let us look at seven of them.

I

The first idea seems very simple at first sight, but it is an entirely negative definition and makes assumptions that we probably cannot justify. It is very much the Cartesian view—that consciousness is the thing we possess that animals lack. Descartes saw the human pineal as a unique route by which something from another plane could drive a flesh-and-blood body. This idea raises the possibility that animals are not conscious, and some would argue that by extension we can treat them as we wish. This may seem instinctively wrong to many readers, but instinct should not drive us here. As usual with consciousness, however, we really have no evidence one way or the other about animal consciousness. There are countless reports of animals interpreting complex situations and formulating impressively insightful plans to deal with them, but can we assume that such quick-wittedness implies consciousness? Also, some thinkers have proposed an anti-Cartesian theory that consciousness is actually very common in nature, and many animals, including animals that seem quite simple, are conscious. Are we so attuned to the idea of using signs of complex thought as evidence of consciousness that we actually have the whole thing the wrong way round? Maybe instead, consciousness is a ubiquitous thing, and a prerequisite for a few rare species to go further and develop intelligence.

II

The second idea of consciousness is not only negative, but also glib. And that is probably why it appeals to me. Just as hackneyed clichés often have a grain of truth within them, maybe this simplest idea of all has a deeper meaning. The theory is that consciousness is the thing that stops when you go to sleep. By "sleep" in this instance, we mean dreamless sleep, as dreaming sleep has some elements similar to wakefulness. Dreaming sleep is probably a relatively recent evolutionary acquisition anyway, as the pattern of alternating episodes of dreaming and dreamless sleep has only been shown to occur in birds and mammals. Although we can be roused from dreamless sleep, the conscious mind does seem somehow suspended. There are many accounts of people walking, talking, writing, murdering, driving, and climbing up cranes to sleep on the jib in their sleep, but for most of us the experience of dreamless sleep is a blissful respite from consciousness. Insomniacs long for their consciousness to go away, even though sleep-deprived people paradoxically undergo relatively less dreamless than dreaming sleep. But the problem is that we do not know what sleep is, or what it is for. There are whole books written about sleep and dreams. Does it permit time for brain repair, or management of the memory filing system, or a way of saving energy by

making animals rest? We simply have no idea. And we are not even
sure that consciousness is suspended in dreamless sleep, or whether
instead the mechanisms that would allow us to recall it are switched
off. Does consciousness count as consciousness if there is no way it
can be remembered?

III

The third idea of consciousness is a good one. It is all about perception and time. If people are asked to say what consciousness is, then
they are more likely to base their answer on how it affects what they
perceive than how it affects what they do. Consciousness can be described as the portal through which we see the universe; the perspective from which our mind engages with the world outside. My world
does not seem to rush past, and it does not seem like a recording. It is
something I am in, and am contemporaneous with. My mind engages
with it even though it is physically lodged inside my skull, and consciousness is what makes that interface so immediate. Sometimes that
interface is temporarily unplugged and I withdraw into my own head,
and this happens most often when I am carrying out a well-worn motor task like playing the guitar or driving to work. I can do both of
these for long periods without apparently "thinking" or paying attention to the world outside. Obviously, my unconscious mind is still doing lots of clever things all that time, but my conscious mind is deliciously rested.

This idea of consciousness as the interface with the world is attractive for all sorts of reasons. First of all, it has a clear evolutionary advantage in that it obviously engages animals with the outside world,
which can only be a good thing. However, equally important is the
fact that it is very selective. When consciousness is not needed it is
"off," and when it is "on," it is a means by which the huge overload of
incoming sensory information is pared down to a relevant core. Some
neuroscientists even think that consciousness "is" that process of
discarding useless information. Whatever the truth, this idea even
gives us potential avenues for study. For example, scientists are actively looking for the brain systems responsible for paying attention—maybe somewhere in the cortex, thalamus, or basal nuclei.

And not only can attentiveness be measured, but conveniently, experimental subjects intermittently lose it spontaneously.

However, it is the idea that consciousness allows us to feel as if we are keeping up with the world outside that has proved most interesting. You probably feel that you are living in the here and now, and you may be surprised to learn that a rich vein of experimental evidence suggests that you are not. Instead, it is the job of consciousness to create an illusion of contemporaneousness. When a sensation hits your eye, ear, or nose, it takes time for the information to reach the brain and then even more time for that signal to be processed and enter consciousness. The amount of time depends on the complexity of the stimulus and varies from a fifth to a half of a second. Because of this, there is a constant risk that simple stimuli may enter the conscious mind before a complex stimulus which preceded them. One of the roles of consciousness is to reorder these perceptions so that they seem to happen in the same order that they occurred in the outside world. Also, it must remove the sense of time lag so that you feel you are aware of stimuli at the instant they enter your body—this may be why direct electrical stimulation of the brain can seem like precognition to experimental subjects. Add to this the likelihood that consciousness must somehow fill in the gaps in your sensory world which occur when you concentrate on something else, and you can see that it has an important and complex editing job to do. It is something that helps us a great deal. Your life is disordered, delayed, and fragmented, but consciousness convinces you that it is a live, smoothly flowing stream.

IV

The fourth concept of consciousness is also a good one. Like the third, it is one that people spontaneously suggest when asked. It is about the self, but not just a crude sense of identity. Instead it is about self-awareness, self-analysis, and internal debate. Many of you will probably have noticed that children gradually develop these abilities as they grow up. I well remember the day that my daughter Rose's friend hurt herself, and although Rose was concerned and worried, her only comment was: "But it does not hurt to *me*." It was an epiphany—she was thinking about pain and how it could take place either within her or outside her. As we mature, this ability gives us two of our most useful abilities—we can empathize with others, and we can improve ourselves by self-criticism. We can look inside ourselves. We are continuously remodeled by our experiences. And conversely we can model possible futures inside our head.

There is a major side-debate about this introspective definition of consciousness, and it relates to language. Some thinkers have defined consciousness in terms of a continual internal verbal discourse—that we all talk to ourselves internally. This is not universally accepted, however, as experiments that require subjects to occasionally report what they are thinking reveal that many of the reported thoughts are not verbal. A requirement for a verbal discussion inside your head would of course have important implications for consciousness in animals, not to mention infants. All jawed vertebrates vocalize, even fish, and some birds and mammals produce apparently complex trains of communicative sounds—songbirds and cetaceans are obvious examples. Yet despite an enormous archive of recorded animal communication, there is no nonhuman animal whose communicative bursts consist of constantly re-ordered sets of relatively few elements—akin to letters, syllables, or words. In contrast, all spoken human languages, no matter how unintelligible, can be demonstrated to conform to exactly that format. We now think that babies are born with a linguistic framework already present in their brains onto which the prevailing language need only be attached, and that this is why all human languages have the same basic pattern. Thus, the language cortex may be a very real novelty in humans, and probably means that humans are the only animals that can articulate novel abstract concepts.

V

The fifth concept of consciousness is really a question rather than a suggestion: does consciousness actually change what we do? This may sound like a silly question and it is tempting to answer in the affirmative, but bear with me. If consciousness does not change what we do, then it is really just a useless side-effect of having a brain. Alternatively, if it does change what we do, then we should be able to detect that change—and at present we cannot. One thing we can do is design experiments in which we ask people to decide to do things. Of course, these people feel that they are making a conscious decision to do something and that this then causes that action to be effected. However, when their brain activity is monitored, something rather frightening becomes evident. Consciousness is editing the world again. Physical movements are usually preceded by an anticipatory flurry of electrical activity in the brain, but experiments suggest that this flurry usually occurs before the time when the subject thought they made the conscious decision to act. There are various interpretations of this finding, but many neuroscientists believe that the anticipatory flurry really does precede the conscious decision. Thus, the conscious mind is not really deciding at all—it is instead feigning decisiveness when some subconscious system has already made the decision for it. If this is the case, then why does consciousness have to trick us into believing that we chose to do something? And where does all this chicanery leave "free will"?

VI

The sixth concept of consciousness is the negative philosophy that consciousness is intrinsically incomprehensible. The reason that it seems so unlike anything else in our experience—uniquely private and familiar yet intangible and indefinable—is that it is indeed unlike anything else in our experience. Having said that, there are several variants of this defeatist approach. One is that an entity (the brain) cannot be intelligent enough to understand itself. This may of course be true, but it can only potentially be demonstrated in the positive. Another suggestion is that, while consciousness is a real phenomenon, the measurable actions of a conscious animal can never be distinguished from those of an animal that acts entirely unconsciously. Whether this is fundamentally true or not, it is certainly true at present, but that should not make us give up.

A further, and very specific, defeatist suggestion is that consciousness is caused by quantum behavior of the elementary particles which make up the brain. I do not claim to understand quantum mechanics, but suffice to say that when the universe is observed at the scale of the very tiny, it behaves in ways that seem entirely unlike the world of our everyday experience. Particles have no defined position or movement; time is a flexible thing; particles wink in and out of existence. Although counterintuitive, quantum mechanics has stood up to experimental testing better than any other theory in the history of science, so the chances that it is wrong are pretty slim. However, whether this apparently chaotic world of the very small is responsible for the capricious, phantom nature of consciousness is far more open to debate.

VII

The seventh and last concept of consciousness is a return to that old idea that our minds are imprisoned in the box in our head. I have already suggested that you have full access to your own consciousness and no access to anyone else's, but is this entirely true? Earlier, I briefly mentioned that social species tend to have larger cerebral cortices, and maybe that statistical fact tells us something. You cannot leap into someone else's consciousness, but is it not the most tantalizing thing to talk with someone and suddenly gain an insight into how they see the world? We are compulsive communicators, our consciousness constantly probing and playing with fellow consciousnesses around it. Our access to other minds is subtle and indirect, but it is that subtlety that makes it so enriching and instructive. Sometimes the things that are partially hidden are the most attractive.

If brains are there to interact with the world outside, then what is the most complex and unpredictable thing in that world, if not the other members of our own species? If your brain is supposed to generate a reflection of the world, then it is creating reflections of beings who are creating their own reflections of you. We are living in a hall of mirrors, all of us assessing our fellows and trying to predict what they want, what they will do, and what they think of us. And they are all doing the same. A social life is far more complicated than an isolated one. So is this why we have consciousness—to create an accessible, sociable façade to interact with others?

Sociability has a long history of making great changes to our bodies. Many vertebrates are sociable, but only mammals evolved a unique way of communicating their state of mind. The muscles that used to move the second gill of our fishy ancestors have swarmed forward onto our face to create muscles of facial expression. This is why dogs can sneer and frown and fawn in a way that budgerigars never can. The second great stage in our social evolution came, of course, when we developed our possibly unique linguistic abilities. Like many human attributes, our language skills have adaptability as their major attribute. We use a small set of noises to communicate instructions, observations, our feelings, and even, heaven forbid, our ruminations on the nature of consciousness. The third step in our sociability was writing—the realization that our verbal communications can be immortalized in graphical symbols. When you read a book, a disembodied person speaks inside your head. Inside your consciousness. Maybe we are not as isolated as we thought.

Having been a teenager in Britain in a decade when the prime minister was spouting such rubbish as "There is no such thing as society," it is wonderful to think that without society there might not be anything at all. Not only no human civilization, but no consciousness with which to mourn its passing. Maybe despite their apparent imprisonment, our consciousnesses are there for each other.

All these ideas of consciousness can flow together and you can take your pick. A thin, ethereal patina of mind above the seething unconscious ferment beneath; a perspective on the world; a way to discard irrelevances; a window on the self; a spur to action; a common ground with our peers.

Still, anathema to fact-lovers like us, it remains without location, rootless. Yet attempts to pick out brain areas that effect consciousness seem faintly ridiculous. The medulla keeps us breathing. The reticular formation, nucleus coeruleus and hypothalamus keep us awake. The sensory association cortex develops our perceptions and puts them in context. The basal nuclei and thalamus maintain our attention. The amygdala emotes. The hippocampus gives us access to the memories that may define our individuality. The prefrontal cortex thinks ahead and spurs to action. And the subcommissural organ may dangle a hairy thing down the middle. All this may be true, but lists like this only serve to emphasize the incompatibility between our view of the brain and our view of consciousness. It is hard to escape the idea that consciousness is somehow special.

But can we move the question of consciousness from the unknowable to the knowable to the known? Perhaps we should use the anatomical approach that has served us so well in the past. If we had a place in our map of the brain for consciousness, then it would all seem so different. Throughout this book we have sought locations for the things that the brain does. A cartography of the *terra incognita* of the mind. As soon as brain functions are found to have a specific location, they immediately become more accessible to us without losing any of their fascination. It is not devaluing the mind to show that it exists in a pale brown mass of fatty strands. Instead, it is ennobling and remarkable. Who would have thought such wonders possible?

EPILOGUE

No Turning Back

They are not long, the weeping and the laughter,
Love and desire and hate:
I think they have no portion in us after
We pass the gate.

They are not long, the days of wine and roses:
Out of a misty dream
Our path emerges for a while, then closes
Within a dream.

—"Vitae Summa Brevis Spem Nos Vetat Incohare
Longam," Ernest Dowson (1867–1900)

All good things must come to an end, and life is one of those things. A while ago, in a billion-to-one coincidence, a particular sperm entered a particular egg and you were conceived. Some time later, your mind flickered into existence. One day it will flicker out of it again.

Death is even more inevitable than taxes. You were never meant to be here for the long haul. Instead, you are a temporary repository for twenty-something thousand genes. If you get lucky, you may have the chance to scatter those genes to the four winds so they can be borne away through future generations. Your brain, your mind, and your consciousness are simply a few more things to help you do that. Those things are here today and gone tomorrow, transient and ephemeral. At some point they must be switched off. How they cope with that switching off is largely irrelevant—what you experience as you die makes no difference to how many babies you make.

But of course, that does not mean that it is not interesting. Would we not all like to know what it feels like to die; whether we slip away peacefully or are torn away in some violent horror? Of course, we can never know for certain what it will be like until it happens, and then we will not be able to tell anybody. However, some people approach death and then turn back. We do not really know how close they get, but they can at least tell us what they experienced. And those near-death experiences are remarkably consistent—so consistent, in fact, that we can perhaps almost expect them. The good news is that apparently most of them are strikingly pleasant.

Key to this pleasantness is a sense of disconnection, or dissociation from the body. Many people who have "come back" tell of a separation of the mind from the body—at least to the point of feeling very passive and accepting of what was happening to their body. A few report an out-of-body experience in which they believe that they were viewing their body remotely, and this can progress to travels to ethereal locations and meetings with supernatural beings, all followed by a recognizable instant when they slot back into their body. Probably the most commonly experienced aspect of dissociation is an incompatibility of one's emotions with what one knows is happening. A few patients report fear, horror, and powerlessness, but if they "give in" to the experience, their feelings change to warm peacefulness, without fear or regret. This state has been likened to the feeling of physical "suspension" before one falls asleep, and in most near-death experiences it is the predominant sensation.

As well as peaceful dissociation, the near-death experience can also induce some more specific experiences. Despite their detailed nature, they are reported so often that people have long wondered what they mean in terms neurological and spiritual. Many reports are based around a journey or a flight along a dark tunnel to a bright white or yellow light. Sometimes this experience of transit is instead a crossing of a river, a walk along a corridor, or the ascent of a flight of stairs, but the inexorable progression into light remains a common theme, memorably depicted in the film of Laura Esquivel's *Like Water for Chocolate*. Of course, by necessity, most of our correspondents report a point at which they turned back from the light, so we can never know how the tunnel ends. Some have even suggested, admittedly with little evidence, that the tunnel is a recollection of the birth canal.

The aspect of the near-death experience that has most fascinated me is often called the "life review." The most familiar form of this is when people in fear of their lives report "my life flashing in front of me." The exact form of the life review can vary. Some describe a rapid chronological recapitulation of their life experiences, while others experience a reverse chronology, with each episode running in the correct direction, but starting with their most recent episodes and working back toward birth. Perhaps most common, however, is the "tableau" format, in which all one's experiences are made simultaneously available—a panorama of your life is spread in front of you for your perusal. Life reviews are, unsurprisingly, entirely personal, but they show some fascinating similarities and differences. They are almost always experienced in the state of warm peacefulness, and often the subject does not apply conceptions of good or bad to what they have done or experienced. But others do acquire a sense of responsibility for their actions as the review progresses, and this may lead to them spending the rest of their life trying to atone for their misdeeds. Conversely, for those who suffer near-death experiences after a suicide attempt, the phenomenon can be a turning point, convincing them that life is something to love rather than dread. Almost all reports of the life review describe the feeling that an immense depth of remembered detail is available—far more than during recollection at other times. However, some report that trivial events are "edited out" of the review whereas others comment that one of its most striking features is that unimportant events are recalled just as well as important ones. And finally, subjects usually mention how leisurely and unhurried the life review seems—appearing to take far longer than their presumed brush with death.

I do not think that near-death experiences are a projection to another plane of existence, and I do not believe that they are an approach toward a supreme being. I simply do not believe in all that hokum. It is a matter of faith and I have no faith in it at all. Instead, I believe that these near-death experiences—the dissociation, the bright light, and the life review—are manifestations of how our minds work. If you accept that all these magnificent phenomena have a physical basis inside our heads, then it simply serves to show what very special creatures we are. Without recourse to the supernatural, the value of human life becomes far greater.

So why does our failing brain undergo near-death experiences? I suppose we should preface this by asking how exceptional these experiences are. First of all, not everyone who comes close to death reports any such experience. Obviously, it is difficult to put numbers on this sort of thing, but many people report no unusual phenomena whatsoever. And conversely, people who believe that they are going to die but who are not thought by their medical caretakers to be in any imminent danger of death still undergo near-death experiences. Although contested, this link of the near-death experience to fear of death rather than its actual likelihood has led some to suggest that these phenomena are the mind's responses to extreme stress rather than evidence that the brain is switching off. Also, none of the main reported features of the experience are entirely unique. We have already seen, for example, that depersonalization disorder can entail a sense of remoteness from the body, as can taking certain drugs, although how these experiences relate to the near-death experience is unclear. Also, the visualization of a tunnel and a bright light can occur in migraine, epilepsy, sleep, and after taking LSD.

I think that this nonspecific nature of the near-death experience is a sign that it may be much more explicable than we thought. One thing that must be borne in mind is that the near-death experience is unlikely to serve any evolutionary purpose. It is difficult to see how it could be advantageous to the individual. The vast majority of animals who experience it must presumably do so as a prelude to death, so it has no conceivable beneficial or adverse effect on their reproductive success. I certainly do not believe that it is some sort of concession from evolution to allow animals comfort as they shuffle off this mortal coil. After all, if the television nature documentaries teach us anything, it is that evolution has produced animals who spend most of their life suffering. Why it should suddenly become kind in our last moments is unclear.

No. The near-death experience is instead a side-effect of the way our brain is put together and how it constructs our consciousness. For example, the dissociation has been suggested to occur as a result of the failure of our sensory association areas. In fact, patients who have in the past reported near-death experiences have different patterns of electrical activity in their temporal lobes, especially, for some reason, on the left. An alternative but equally practical suggestion is that dissociation is a normal protective response to extreme stress—taking you away from things that you simply cannot bear, perhaps by releasing a huge burst of calming chemicals into the brain. The bright light at the end of the tunnel has for some time been suspected to result from oxygen starvation of the visual system—either the retina of the eye, or more probably the visual processing regions of the cortex. Maybe it has something to do with the fact that we only see the center of the visual field clearly anyway, and the relatively small amount of information we gather from the periphery is more easily lost. Do dying dogs experience "tunnel smell"?

The life review is more difficult to explain, but we should not take that to mean that it cannot be explained. Why should we suddenly be presented with a potted history of our life in such an apparently leisurely fashion? I propose that the life review is a reflection of how our brain stores memories and keeps time, and how these systems behave as they deteriorate during death. Memory is usually an active thing— something brings memories back into our consciousness, whether bidden or unbidden. Although memories are a large part of what we are as individuals, we are usually unaware of their profusion or organization. We have ways of drawing them back to mind and filing them away again, but we are hidden from the sheer fright of seeing what an intimidatingly large memory bank we possess. Perhaps it is something best kept from us. I would suggest that the life review occurs when the expiring brain loses this active selectivity, and our memories all become simultaneously and equally accessible. I cannot explain why some people experience a chronology, others a reverse chronology, and others a tableau, but perhaps you can think of some reasons.

The leisureliness of the review is probably a sign that the brain's internal timekeeping system is failing. We now think that there are circuits in the forebrain which tick out our subjective view of time. This timekeeping can vary during our normal activities, which is why time can seem to pass slowly or quickly depending on whether you are bored, excited, unhappy, or joyous. It is not too much to expect that this timekeeping could go seriously askew as the brain begins to shut down.

The near-death experience does not achieve anything, but that does not mean that it is not important to us. Evolution gave us a brain and that brain has constructed for each of us a consciousness with which to deal with our world. It is the immersing, contemporary interface with that outside world and yet it is an internal, personal experience. The brain is what allows us to exist in the world. Surely we would all like to know how it feels when it is time to leave?

FURTHER READING

1: Skull Marrow

The original, and still definitive translation of the Edwin Smith surgical papyrus is still

J. H. Breasted, *The Edwin Smith Surgical Papyrus: Published in Facsimile and Hieroglyphic Transliteration with Translation and Commentary* (Chicago, IL: University of Chicago Press, 1930).

A variety of ancient writings on the nervous system may be found in the following

E. Clarke and C. D. O'Malley, *The Human Brain and Spinal Cord: A Historical Study Illustrated by Writings from Antiquity to the Twentieth Century* (Los Angeles, CA: University of California Press, 1995).

2: Servants and Guards of the Great King

Good introductions to ancient Greek thinking on the brain and the work of Galen himself:

Aristotle, trans. W. Ogle, *De partibus animalium* (London: K. Paul, French & Co, 1882);
Galen, trans. M. T. May, *On the Usefulness of Parts of the Body* (Ithaca, NY: Cornell University Press, 1968);
J. Rocca, *Galen on the Brain* (Leiden: Brill, 2003).

And trepanation is still with us:

www.trepan.com.

3: The Brain as Geography

Geographical parallels with the naming and misnaming of the brain abound in both

M. Jancey, *Mappa Mundi: The Map of the World in Hereford Cathedral* (Hereford: Hereford Cathedral Enterprises, 1994);
D. S. Johnson, *Phantom Islands of the Atlantic* (London: Souvenir, 1997).

4: A River Runs Through It

A general introduction to embryonic development and a discussion of the folding-in of the brain tube:

D. R. J. Bainbridge, *Making Babies: The Science of Pregnancy* (Cambridge, MA: Harvard University Press, 2000);
J. L. Smith and G. C. Schoenwolf, *Trends in Neuroscience* 20 (1997): 510–517.

How the brain "inflates" rather than "grows":

G. C. Schoenwolf and M. E. Desmond, *Journal of Experimental Zoology* 230 (1984): 405–407.

And how it can all go wrong:

E. R. Detrait et al., *Neurotoxicology and Teratology* 27 (2005): 515–524. (review of neural tube defects).

The embryo images were all derived from a book by my favorite nineteenth-century scientist:

E. F. A. Haeckel, *Anthropogenie, oder Entwickelungsgeschichte des Menschen* (Leipzig: Wilhelm Engelmann, 1877).

Why leaking cerebrospinal fluid from your nose is not a good thing:

T. Okuda, K. Kataoka, M. Kitano, A. Watanabe, and M. Taneda, *Minimally Invasive Neurosurgery* 48 (2005): 247–249.

A case of fetus-in-the-brain:

D. L. Kimmel, E. K. Moyer, A. R. Peale, L. W. Winborne, and J. E.
   Gotwalss, *Anatomical Record* 106 (1950): 141–165.

5: Leonardo's Butterfly

Some scientific papers on the causes and treatment of multiple sclerosis:

J. Clausen, *International Multiple Sclerosis Journal* 10 (2003): 22–28 ;
B. Hemmer, O. Stuve, B. Keiseier, H. Schellekens, and H. P. Hartung,
   Lancet Neurology 4 (2005): 403–412;
L. E. Hughes et al., *Journal of Neuroimmunology* 144 (2003): 105–115.

6: Interlude

Some recent discussions of the vertebrate-as-inverted-fly controversy:

E. M. DeRobertis and Y. Sasai, *Nature* 380 (1996): 37–40;
S. A. Holley and E. L. Ferguson, *BioEssays* 19 (1997): 281–284.

7: A Forest So Dense

I thoroughly recommend Santiago Ramón y Cajal's humorous and endearing autobiography, originally published as:

S. Ramón y Cajal, *Recuerdos de mi Vida* (Madrid: Juan Pueyo, 1923).

8: The Little Fish Who Never Grew Up

The literature on the evolution and mechanisms of hearing is enormous, but you could try:

E. Borg and J.-E. Zakrisson, "The Stapedius Muscle and Speech Perception" in: *Sound Reception in Mammals*, ed. R. J. Bench, A. Pye, and
   J. D. Pye (London: Academic Press, 1975);
A. N. Popper and R. R. Fay, *Brain, Behaviour and Evolution* 50 (1997):
   213–221.

9: The Brain as Archaeology

The involvement of the hindbrain in abnormal breathing patterns and spongiform encephalopathies:

C. L. Bassett and M. Gugger, *Swiss Medical Weekly* 132 (2002): 109–118;

R. Klitzman, *The Trembling Mountain: A Personal Account of Kuru, Cannibals, and Mad Cow Disease* (New York: Perseus Publishing, 2001);
C. Weissmann, *Journal of Biological Chemistry* 274 (1999): 3–6.

10: Beauty Is in the Eye of the, Er, Squid

There is a great deal more about color vision and the difference between the sexes in my last book:

D. R. J. Bainbridge, *The X in Sex* (Cambridge, MA: Harvard University Press, 2002).

11: Hillocks, Buttocks, Blindsight, and Black Stuff

The fascinating phenomenon of "blindsight":

C. E. Collins et al., *Proceedings of the National Academy of Sciences (USA)* 102 (2005): 5594–5599;
B. J. Liddell et al., *Neuroimage* 24 (2005): 235–243.

There is an enormous literature on Parkinson's disease, but articles on the strange role of the brain's black pigments and the use of embryonic cells to treat the disease include:

H. Fedorow, F. Tribl, G. Halliday, M. Gerlach, P. Reiderer, and K. L. Double. *Progress in Neurobiology* 75 (2005): 109–124;
S. U. Kim, *Neuropathology* 24 (2004): 159–171.

12: Stinkin' and Thinkin'

For a strange and diverse sense, smell has an appropriately strange and diverse literature. For more on smell evolution, the organ of Jacobson, and incest-avoidance, try:

S. Rouquier, A. Blancher, and D. Giorgi, *Proceedings of the National Academy of Science U.S.A.* 97 (2000): 2870–2874;
L. Watson, *Jacobson's Organ: And the Remarkable Nature of Smell* (New York: Plume, 2001);
G. E. Weisfeld, T. Czilli, K. A. Phillips, J. A. Gall, and C. M. Lichtman, *Journal of Experimental Child Psychology* 85 (2003): 279–295.

13: Into the Marriage Chamber for Some Sexy Synesthesia

For more on the strange, wonky structures in the roof of the third ventricle and their possible roles in addiction:

M. L. Concha and S. W. Wilson, *Journal of Anatomy* 199 (2001): 63–84;
G. Ellison, *European Neuropsychopharmacology* 12 (2002): 287–297.

14: Why Is "D" Brown?

There is an extensive literature on the wild world of synesthesia. These three are, in turn, a beautiful book on synesthetic music, a review of the neural mechanisms of synesthesia, and an investigation of why synesthesia and number forms are often based on ordered sequences.

K. Brougher, J. Strick, A. Wieman, and J. Zinczer, *Visual Music* (London: Thames and Hudson, 2001);
A. N. Rich and J. B. Mattingley, *Nature Reviews Neuroscience* 3 (2002): 43–53;
N. Sagiv, J. Simner, J. Collins, B. Butterworth, and J. Ward, *Cognition* 7 (2005): e-publication: no page numbers.

15: Interlude

The "Nick et al." study in which I was a subject has the wonderful name "Depersonalization Disorder: Thinking without Feeling":

M. L. Phillips, N. C. Medford, C. Senior, E. T. Bullmore, J. Suckling, M. J. Brammer, C. Andrew, M. Sierra, S. C. Williams, and A. S. David, *Psychiatry Research* 108 (2001): 145–160.

16: The Brain as Engineering

Penfield published several eloquent descriptions of his work on "mapping" the cortex as well as an autobiography. These include:

W. G. Penfield, *No Man Alone: A Surgeon's Life* (New York: Little, Brown, 1977);
———, *The Mystery of Mind* (Princeton: Princeton University Press, 1975);
W. G. Penfield and T. Rasmussen, *The Cerebral Cortex of Man* (New York: Macmillan, 1957).

17: The Apparent Disorder of the Cerebral Jungle

Of course, there is a profuse and ever-growing literature on the cortex, but a few that I used include:

M. Behrmann, J. J. Beng, and S. Shomstein, *Current Opinion in Neurobiology* 14 (2004): 212–217;

L. Fogassi and G. Luppino, *Current Opinion in Neurobiology* 15 (2005): 626–631;

M. L. Kringelbach, *Nature Reviews: Neuroscience* 6 (2005): 691–792.

18: The Seahorse and the Almond

The dark, fearful world of the amygdala and the organized filing cabinet of the hippocampus:

C. Broglio et al., *Brain Research Bulletin* 66 (2005): 277–281;

S. Hamann, *Neuroscientist* 11 (2005): 288–293;

E. L. Kier, J. H. Kim, R. K. Fulbright, and R. A. Bronen, *American Journal of Neuroradiology* 18 (1997): 525–532;

P. M. Maki, *International Journal of Fertility and Women's Medicine* 50 (2005): 67–71.

19: The Hard Question

Brain size, human brain growth, and the "hard question" of consciousness:

A. B. Butler, P. R. Manger, B. I. Lindahl, and P. Arhem, *Bioessays* 27 (2005): 923–936;

L. Lefebvre, S. M. Reader, and D. Sol, *Brain Behaviour and Ecology* 63 (2004): 233–246;

M. Minsky, *The Society of Mind* (London: Heinemann, 1985);

C. Ponting and A. P. Jackson, *Current Opinion in Genetics and Development* 15 (2005): 241–248.

Epilogue

D. R. J. Bainbridge, *Fortean Times* 162 (2002): 53–54;

J. E. Owens, E. W. Cook, and I. Stevenson, *Lancet* 336 (1990) 1175–1177;

H. Thurston and R. Rickard, *Fortean Times* 159 (2002): 34–38.

Finally I would like to acknowledge the fount of knowledge that is the fellowship of St. Catharine's College, Cambridge. Where else could one clarify the erroneous etymology of the word "petalia" and be informed of the number of cochlear turns in a capybara by return of e-mail on a national holiday?

INDEX